Lecture Notes in Physics

Bisher erschienen/Already published

Lecture Notes in Physics

Edited by J. Ehlers, München, K. Hepp, Zürich
R. Kippenhahn, München, H. A. Weidenmüller, Heidelberg
and J. Zittartz, Köln
Managing Editor: W. Beiglböck, Heidelberg

74

Pierre Collet
Jean-Pierre Eckmann

A Renormalization Group Analysis of the Hierarchical Model in Statistical Mechanics

Springer-Verlag
Berlin Heidelberg GmbH 1978

Authors
Pierre Collet
Jean-Pierre Eckmann
Département de Physique Théorique
Université de Genève
32, Boulevard d'Yvoy
1211 Genève 4/Switzerland

ISBN 978-3-540-08670-3   ISBN 978-3-540-35899-2 (eBook)
DOI 10.1007/978-3-540-35899-2

© by Springer-Verlag Berlin Heidelberg 1978
Originally published by Springer-Verlag Berlin Heidelberg New York in 1978

2153/3140-543210

TABLE OF CONTENTS

INTRODUCTION

The so-called renormalization group (RG) theory which has seen a vigorous development in the past few years has considerably strengthened our understanding of phenomena near to phase transitions of statistical mechanics, and it has also given some insight into the difficulties of relativistic quantum field theories. Maybe the main virtue of the RG theory has been to ask the right questions, namely to put the study of collective phenomena (that is the cooperative behaviour of many particles or modes) into a good perspective. The method consists of studying the behaviour of a physical system under a change of scale. The study of this question can be separated into two parts :

Firstly, to ask in which way the microscopic physical laws transform under such a change of scale, and secondly, to ask why and how information about the system near a "critical" situation can be obtained once the transformations of these microscopic laws are known.

The second question has been essentially completely answered in the literature on critical phenomena while the first still poses some interesting problems. In these Lecture Notes we address ourselves exclusively to the second question by considering a model (the Hierarchical Model) in which the first problem is completely answered by construction. This approach is then sufficiently modest to allow for a complete mathematical understanding of the following main problems of RG theory which are : The existence of non-trivial fixed points, their ε-expansion, local flows and crossover phenomena and the physical information which can be extracted from these things.

These mathematical problems have been first solved by Bleher and Sinai and most of the proofs can be found in the references by these authors. The present Lecture Notes report these ideas in our realiza-

tion with proofs which differ sometimes essentially from those of
Bleher and Sinai.The study of the ε-expansion follows our own earlier
work, while the existence proof given here is new and our crossover
proofs are more detailed than those of Bleher and Sinai.

These Lecture Notes are written in two parts which are distinct
in style. In Part I we develop the different aspects of the renormalization group for the Hierarchical Model. These aspects are mostly
given in the form of a more intuitive exposition followed by a precise
mathematical statement. Those calculations which seem instructive are
given in Part I but only the strategy of the proofs is outlined. Our
approach to the subject is not along the conventional line because it
is exclusively based on statistical mechanics, i.e. thermodynamic
quantities appear as derived objects. It may be useful to read one of
the review articles by Ma[1], Wilson-Kogut [2] or Fisher [3], to make
contact with the more thermodynamic approach.

Part II serves a different purpose : It is a complete mathematical
description of all steps used in the arguments of Part I. Many of the
results were shown before by Bleher and Sinai and are scattered in the
literature. Our proofs are however new and many of them appear here
for the first time. The language is that of mathematics and we address
readers familiar with functional analysis.

REFERENCES

[1]   Sh. K. MA   Introduction to the renormalization group.
Rev. Mod. Phys. 45, 589 (1973).

[2]   K.G. WILSON, I. KOGUT   The renormalization group. Phys. Rep.
12C, 75 (1974).

[3]   M.E. FISHER   The renormalization group in the theory of critical behaviour. Rev. Mod. Phys. 46, 597 (1974).

PART I. HEURISTICS

1. Probabilistic Formulation of the Problem

The success of the RG method rests in part on the fact that statements are only made about a very restricted number of observables of a system. Most of these observables describe the collective behaviour of many degrees of freedom. Typical such observables describe the value of the mean spin of a system, or the fluctuations of this mean.

Probability theory asks similar questions : Given random variables s_1, ..., s_n, with probability densities $\rho_1(s_1) = \rho(s_1)$, we may ask for the probability density of the sum (or the mean) of the s_i . The answer is well known ; the probability density P_N for $S = s_1 + \ldots + s_N$ is for independent random variables,

$$P_N(S) = \int ds_1 \ldots ds_N \ \rho(s_1) \ldots \rho(s_N) \ \delta(S - s_1 - \ldots - s_N).$$
$$(1.1)$$

How does P_N behave in the limit of large N ? The central limit theorem answers this question.

THEOREM 1.1.[*] Let $\mu = \int s\rho(s) \ ds < \infty$, $\sigma^2 = \int (s - \mu)^2 \rho(s)ds < \infty$. Then

$$\lim_{N \to \infty} P_N (N^{\frac{1}{2}} (S + N \mu)) = (2\pi \ \sigma^2)^{-\frac{1}{2}} e^{-S^2/2\sigma^2},$$
$$(1.2)$$

where the convergence is in the weak sense $\Big($i.e.

$$\int f(S) P_N (N^{\frac{1}{2}} (S + N \mu)) \ dS \quad \to \int (2\pi \ \sigma^2)^{-\frac{1}{2}} f(s) \ e^{-s^2/2\sigma^2} \ ds$$

for $f \in L_1(dS)\Big)$.

[*] The notation A/BC means always A/(BC) .

The formulae (1.1), (1.2) exhibit many typical features of RG theory : indeed, a question is asked about a sum of variables(sum of spins). This sum is rescaled (renormalized) through $(s-N\mu)/N^{\frac{1}{2}}$. Most importantly the limit has a behaviour which is <u>independent</u> of the details of ρ, and the analogous feature in RG theory is called universality.

The situation described in (1.1), (1.2) corresponds to a free classical discrete spin system with continuous spin (a value in R), and this can be seen as follows. Choose $\beta > 0$ (the inverse temperature) and set $H(s) = - (\log \rho(s))/\beta$, $H_N(s_1, \ldots, s_N) = \sum_{j=1}^{N} H(s_j)$. $H(s_j)$ is the "energy" of the spin s_j . The expectation for the fluctuation of the sum of the spins $S_N = \sum_{j=1}^{N} s_j$ is then given by

$$\chi_N^2 = < (S_N - < S_N >)^2 > / N , \qquad (1.3)$$

where

$$< f(s_1, \ldots, s_N) > = \frac{\int ds_1 \ldots \int ds_N \ e^{-\beta H_N(s_1, \ldots, s_N)} f(s_1, \ldots, s_N)}{\int ds_1 \ldots \int ds_N \ e^{-\beta H_N(s_1, \ldots, s_N)}}$$

$$(1.4)$$

is the expectation of f in the Gibbs ensemble of statistical mechanics.

Theorem 1.1. implies by inspection that as $N \to \infty$,

$$\chi_N^2 \quad \to \sigma^2 . \qquad (1.5)$$

Thus the fluctuations in any free spin system for which a single spin has finite mean μ and variance σ behave asymptotically like $N^{\frac{1}{2}}\sigma$ as a function of the number N of particles.

In the course of the study of the model, we shall not only concentrate on fixed points but also on the "flow" around them, i.e. on
the approach to the fixed points. In fact, from a physical point of
view, the latter problem is more important than the former, because
it allows to make statements about large but finite systems.

As in probability theory, one can ask which distributions φ can
occur as limits of initial distributions under some transformations.
This is a deep problem, which is completely solved in the case of independent random variables. Also the domain of attraction (= universality class) of each possible limit distribution (which are called the
stable distributions in the mathematics literature) is known in this
case ; i.e. one can say which distributions"converge" to which limits.
Some attempts to make progress in this difficult problem for dependent
variables have been made by Sinai[8] and Bleher but they have not yet
gone beyond some beautiful but straightforward generalization of phenomena which will show up already in the study of the special case of
the hierarchical model. However, the benefit of the probabilistic
description of the RG has certainly been to put the notion of universality classes into precise language.

Remarks on Section 1 :

The probabilistic interpretation of the RG has been stressed especially by Jona-Lasinio. Earlier allusions are made in passing in Bleher-
Sinai, Baker.

[4]   G. JONA-LASINIO : The renormalization group : A probabilistic
                view. Il Nuovo Cimento 26B, 99 (1975).

[5]   P.M. BLEHER,Ja.G. SINAI : Investigation of the critical point
                in models of the type of Dyson's Hierarchical
                Model. Commun.Math. Phys. 33, 23 (1973).

[6]   G.A. BAKER :   Ising model with a scaling interaction. Phys.
                Rev. B 5,2622 (1972).

A detailed study of sequences of independent random variables can be
found in

[7]   B.V. GNEDENKO, A.N. KOLMOGOROV : Limit distributions for sums
                of independent random variables. Cambridge
                Mass. 1954, Addison Wesley.

[8]   Ja.G. SINAI :   Self-similar probability distributions. Theory
                of probability and its applications 21, 64
                (1976).

2. The RG-Transformation for the Hierarchical Model

We start this section by defining the model. The Hierarchical Model is a model of continuous spins on a one-dimensional lattice. If the lattice has N points, the spins will be called $s_1, \ldots, s_N$. For every real function f and every $N = 2^M$ we define the Hamiltonian $\mathcal{H}_{N,f}$ of the system to be

$$\mathcal{H}_{N,f} = \mathcal{H}_N + \sum_{j=1}^{N} f(s_j), \tag{2.1}$$

$$\mathcal{H}_N = -\sum_{k=1}^{M} \frac{c^k}{2^{2k+1}} \sum_{j=0}^{2^{M-k}-1} \left(\sum_{l=1}^{2^k} s_{j2^k+l} \right)^2. \tag{2.2}$$

The constant c is real and $1 < c < 2$.

We do not discuss at this point for which values of c and for which choices of f the Hamiltonian actually defines a thermodynamically stable system. Let us now describe the heuristics of Eq. (2.2). The Hamiltonian $\mathcal{H}_N$ is the sum of terms on "levels" $k=1, \ldots$ M. On each level k, the 2^M spins are grouped into disjoint blocks of 2^k spins each and the interaction for such a block is then

$$-\frac{c^k}{2^{2k+1}} \left(\sum_{l=1}^{2^k} s_{j2^k+l} \right)^2 .$$

This is usually visualized graphically as follows:

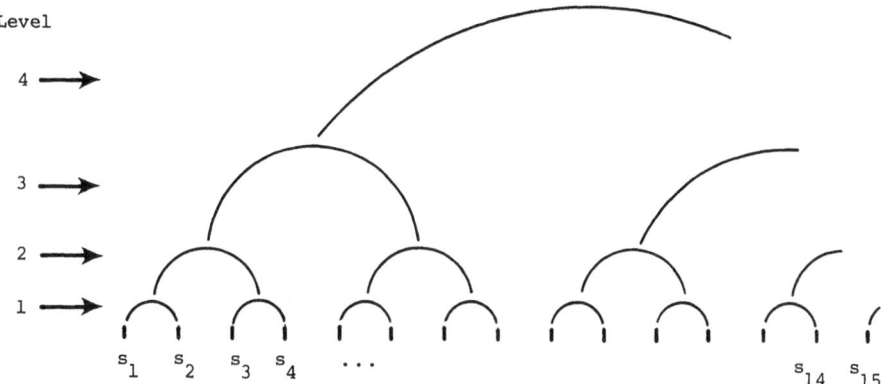

Figure 1. The hierarchical structure of the interaction

Let us study the interaction between s_i and s_j, $i \neq j$. By the nature of the Hamiltonian, there will be a lowest level for which s_i and s_j lie in the same block, say the level k. Then the interaction between s_i and s_j is $-(c/4)^k$. On the other hand, the fact that the lowest level is k implies $|i-j| \geq 1$ and $|i-j| \leq 2^k-1$. It is thus reasonable to say that the interaction potential is about of the form $|i-j|^{\log_2(c/4)}$ but this is to be taken with a grain of salt because the model is not translation invariant. We thus see that the range of the interaction depends on c.

Most often RG theory is done in varying dimension for short range interactions. In the case of the Hierarchical Model, the situation is reversed in that the dimension is fixed and the range of the interaction is varied. While this is unusual, it has the advantage of being more easily implementable from a mathematical point of view than the notion of fractional space dimension.

Let us now assume $f(s)$ is sufficiently increasing at infinity so that $\int \prod_{i=1}^{N} ds_i s_i^{k_i} \exp(-\beta \mathcal{H}_{N,f}(s))$ exists. Then the model is defined for all finite volumes and we may discuss its partition function. In particular, we shall consider the probability density for the sum of spins, as in Section 1, at inverse temperature $\beta > 0$. It is

$$P_{N,f}^{(\beta)}(S) = \frac{\int ds_1 \ldots ds_N \; \delta(S - s_1 - \ldots - s_N) e^{-\beta \mathcal{H}_{N,f}}}{\int ds_1 \ldots ds_N \; e^{-\beta \mathcal{H}_{N,f}}} \qquad (2.3)$$

We shall now compare $P_{2N,f}^{(\beta)}$ and $P_{N,g}^{(\beta)}$, using the explicit definition (2.3) and the special form of the Hamiltonian (2.2). Observe that for $k \geqslant 1$,

$$\sum_{l=1}^{2^k} s_{j2^k+l} = \sum_{l=1}^{2^{k-1}} s_{j2^k+2\,l-1} + s_{j2^k+2\,l} \; .$$

Therefore, for $N = 2^M$, $M \geqslant 0$, we find

$$\mathcal{H}_{2N}(s_1, \ldots s_{2N})$$

$$= - \sum_{k=1}^{M+1} \frac{c^{k-1}}{2^{2(k-1)+1}} \sum_{j=0}^{2^{M+1-k}-1} \left(\sum_{l=1}^{2^{k-1}} \frac{c^{\frac{1}{2}}}{2} (s_{j2^k+2\,l-1} + s_{j2^k+2\,l}) \right)^2$$

$$= \mathcal{H}_N\left((s_1+s_2) c^{\frac{1}{2}}/2, (s_3+s_4) c^{\frac{1}{2}}/2, \ldots, (s_{2N-1} + s_{2N}) c^{\frac{1}{2}}/2 \right)$$

$$- \frac{c}{8} \sum_{j=0}^{2^M-1} (s_{2j+1} + s_{2j+2})^2 \; .$$

Therefore we find, for any measurable function F,

$$\int ds_1 \ldots ds_{2N} \; F\left(\sum_{j=1}^{2N} s_j\right) \; \exp(-\beta\mathcal{H}_{2N,f}\,(s_1, \ldots, s_{2N}))$$

$$= \int ds_1 \ldots ds_{2N} \; \prod_{j=1}^{2N} \exp(-\beta f(s_j)) \; \prod_{j=1}^{N} \exp(\beta c(s_{2j-1} + s_{2j})^2/8)$$

$$\cdot F\left(\sum_{j=1}^{2N} s_j\right) \; \exp(-\beta\mathcal{H}_N(\,(s_1 + s_2)\,c^{\frac{1}{2}}/2, \; \ldots, \; (s_{2N-1} + s_{2N})c^{\frac{1}{2}}/2)),$$

which, upon setting

$$t_j = (s_{2j-1} + s_{2j})c^{\frac{1}{2}}/2 \;, \qquad u_j = (s_{2j-1} - s_{2j}) /2,$$

becomes

$$(2/c^{\frac{1}{2}})^N \int dt_1 \ldots dt_N \; F\!\left(2c^{-\frac{1}{2}} \sum_{j=1}^{N} t_j\right) \exp(-\beta\mathcal{H}_N(t_1, \ldots t_N))$$

$$\cdot \prod_{j=1}^{N} \int du_j \; \exp(-\beta f\,(t_j c^{-\frac{1}{2}} + u_j) -\beta f\,(t_j c^{-\frac{1}{2}} - u_j)) \; \exp(\beta\, t_j^{\,2}/2)$$

$$= \int dt_1 \ldots dt_N \; F\!\left(2c^{-\frac{1}{2}} \sum_{j=1}^{N} t_j\right) \exp(-\beta\mathcal{H}_{N,g}\,(t_1, \ldots, t_N)),$$

where g is defined by

$$e^{-\beta g(t)} = e^{\beta t^2/2} \, (2/c^{\frac{1}{2}}) \int du \; e^{-\beta\,f(tc^{-\frac{1}{2}} + u) - \beta\,f(tc^{-\frac{1}{2}} - u)} \;.$$

$$(2.4)$$

The Equation (2.4) defines a transformation

$$f \longrightarrow g = \mathcal{N}_p^{(\beta)}(f) \;,$$

and upon inserting our calculations into (2.3) we find the important relation

$$P_{2N,f}^{(\beta)}(S) = (c^{\frac{1}{2}}/2) \; P_{N,\mathcal{N}_p}^{(\beta)}(\beta)(f) \quad ((c^{\frac{1}{2}}/2) \; S) \quad .$$

$$(2.4a)$$

What have we now achieved ? We have related the probability densities corresponding to two different numbers of spins (namely N and 2N)
through a change of scale (1 goes to $2/c^{\frac{1}{2}}$) and by a change of Hamilto-
nian $\mathcal{H}_{.,f} \longrightarrow \mathcal{H}_{.,\mathcal{N}_p}(\beta)(f)$. Putting it slightly differently : A simultaneous change of scale and of the Hamiltonian has no effect. The
semigroup formed by these simultaneous transformations is called the
renormalization group. The Hierarchical Model is an especially simple
system insofar as the change of Hamiltonian concerns only the single
spin distribution f . In the general framework of the renormalization
group theory the transformation of the Hamiltonian involves other terms,
too. The simple structure of the RG transformation for the Hierarchical
Model will make a rigorous mathematical discussion possible, while the
typical features of RG theory are preserved.

What can these RG equations be used for ? First of all we recast
them into a form which shows the similarities with the probabilistic
aspects discussed in Section 1. The quantity P_N which we considered
there satisfied the equation

$$P_{2N}(S) = \int dT \; P_N(S/2-T) \; P_N(S/2+T),$$

and the central limit theorem (Theorem 1.1) asserted (in the case of
zero mean $\mu = 0$)

$$\lim_{M \to \infty} P_{2^M} (2^{M/2} S) \longrightarrow Gaussian .$$

$$(2.5)$$

For $N = 2^M$ we may also decompose the Hamiltonian as the following sum :

$$\mathcal{H}_{2N,f} (s_1,\ldots,s_{2N}) = \mathcal{H}_{N,f} (s_1,\ldots,s_N) + \mathcal{H}_{N,f} (s_{N+1},\ldots s_{2N}) .$$

$$- \tfrac{1}{2} c^{M+1} 2^{-2M-2} (\sum_{j=1}^{2N} s_j)^2 .$$

Then by a sequence of transformations similar to those leading to (2.4) we get

$$P_{2N,f}^{(\beta)} (S) = \text{const. } \exp(\beta \ c^{M+1} \ 2^{-2M-2} \ S^2/2)$$

$$\cdot \int dT \ P_{N,f}^{(\beta)} (S/2 - T) \ P_{N,f}^{(\beta)} (S/2 + T) .$$

In analogy with (2.5) we may consider

$$P_{2^M,f}^{(\beta)} \left((2/c^{\tfrac{1}{2}})^M \ S \right) = \mathcal{R}_{2^M,f}^{(\beta)} (S),$$

which then satisfies

$$\mathcal{R}_{2^{M+1},f}^{(\beta)} (S) = \text{const. } \exp(\beta \ S^2/2)$$

$$\cdot \int du \ \mathcal{R}_{2^M,f}^{(\beta)} (S \ c^{-\tfrac{1}{2}} + u) \ \mathcal{R}_{2^M,f}^{(\beta)} (S \ c^{-\tfrac{1}{2}} - u), \qquad (2.6)$$

as compared to $\mathcal{R}_{2^M}(S) = P_{2^M} (2^{M/2} S)$, in the case discussed in Section 1 which satisfies

$$\mathcal{R}_{2^{M+1}} (S) = \text{const.} \int du \ \mathcal{R}_{2^M} (S \ 2^{-\tfrac{1}{2}} + u) \ \mathcal{R}_{2^M} (S \ 2^{-\tfrac{1}{2}} - u) .$$

$$(2.7)$$

The equation (2.6) is very similar to (2.7) which we discussed in Section 1. But the very regular situation described in Equ. (2.5) may now change drastically for one value of β, called the _inverse critical temperature_. Then the fluctuations can be for example of order $N^\tau, \tau \neq 1$, and S could tend to Gaussian distribution with variance $N^{\tau/2}\sigma$;

$$X_N^2 / N^{\tau-1} \rightarrow \sigma^2, \qquad \tau \neq 1 . \qquad (2.8)$$

(In our case $\tau = 2 - \log_2 c$).

Finally, there is the possibility that $\tau \neq 1$ and in addition $S_N/N^{\tau/2}$ does not tend to a Gaussian distribution, but to some other distribution Φ . This third case

$$P_{N,f}^{(\beta)} \left(N^{\tau/2} (S + N \mu) \right) \rightarrow \Phi (S) \neq (2\pi\sigma^2)^{-\frac{1}{2}} e^{-S^2/2\sigma^2} , \qquad (2.9)$$

is the most interesting one from a physical point of view, and the limit Φ is called a _nontrivial critical spin distribution_; or (the exponential of) a critical Hamiltonian. We prefer the first interpretation, and this is the reason for having exposed the RG in the probabilistic framework . (In mathematics Φ would correspond to the distribution of a sum of _dependent_ random variables.) We shall see that in the Hierarchical Model behaviour of the type Eq.(2.9) occurs. The purpose of these Lecture Notes is among others to study this generalized form of a central limit theorem for the Equation (2.6). But we _view the limit_ itself _as a fixed point of the transformation_

$R_{N,f}^{(\beta)} \rightarrow R_{N+1,f}^{(\beta)}$ defined by Equation (2.6). In fact, we shall not work with (2.6), which we used to show the connection between the RG theory and the central limit theorem, but we shall rather concentrate on the transformation $\mathcal{N}_P^{(\beta)}$ defined in (2.4), which also describes the scaling behaviour of the main object, namely $P_{N,f}^{(\beta)}$, which is defined in Eq. (2.3).

These Lecture Notes are then a <u>study of the transformation</u> $\mathcal{N}_p^{(\beta)}$. Two main methods for this study are used :

M 1) Look for a fixed point of the map $\mathcal{N}_p^{(\beta)}$. Then under suitable conditions, the behaviour of the map in a neighborhood of the fixed point is completely described in terms of the tangent map at the fixed point. We shall see later that $\mathcal{N}_p^{(\beta)}$ has fixed points which are not Gaussian, and these are the ones of special interest to us.

M 2) Follow trajectories globally. This method is much less systematic than the first one and our results are maybe mathematically not so appealing.

The above methods allow both for strong results about the system. From a physical point of view the results provided through M 1 and M 2 are distinct.

M 1 allows to determine the <u>critical indices,</u> i.e. to determine the behaviour of thermodynamic variables when the temperature reaches the critical temperature. M 1 corresponds to the so-called scaling limit. The fact that the result is independent of some class of functions f reflects what is called the universality character of the RG method.

M 2 allows to prove, for suitable functions f in $\mathcal{K}_{N,f}$, and for suitable observables, the <u>existence of the thermodynamic limit,</u> i.e. the limit $M \to \infty$ in (2.6), at temperatures near, but not equal to a specific temperature, called the critical temperature. In addition it implies that the mean spin and the correlation length are finite when the temperature is not critical. Finally, the existence of a phase transition at the critical temperature follows. (Such results can often be obtained by totally different arguments, but the RG treatment seems particularly nice in the context of the Hierarchical Model. Furthermore the results on finite correlation length outside the critical temperature are not known except for the Ising model).

As we have seen above, the Hierarchical Model has the property that its RG transformation $\mathcal{N}_P^{(\beta)}$ is a known transformation on the space of __single__ spin distribution. This is not the case for a general model, but believed to be approximately true for large N. Whenever this should be the case for a transformation sufficiently similar to $\mathcal{N}_P^{(\beta)}$ (e.g. convolution of several factors and a Gaussian factor) the ideas of these Lecture Notes could be carried over. However, the determination of a "correct" approximate RG transformation is a very hard problem for a general microscopic Hamiltonian, and we do not pursue this question any further.

In the next section, we shall discuss the existence of a nontrivial fixed point of the transformation $\mathcal{N}_P^{(\beta)}$, and we shall come back to the application of Method M 1 in later sections.

Remarks on Section 2 :

The Hierarchical Model has been invented by Dyson to show that one-dimensional systems may exhibit phase transitions if they have long-range forces.

[9]   F.J.DYSON : Existence of a phase-transition in a one-dimensional Ising ferromagnet . Commun. Math.Phys. 12, 91 (1969).

[10]   F.J.DYSON : An Ising ferromagnet with discontinuous long-range order. Commun. Math. Phys. 21, 269 (1971).

Baker reinvented the model and pointed out that the RG acted on the single spin distribution. He also calculated critical indices.

[11]   G.A. BAKER, Jr : Ising model with a scaling interaction. Phys. Rev. B5 , 2622,(1972).

[12]   G.A. BAKER, Jr ; G.R. GOLNER : Spin-spin correlations in an Ising model for which scaling is exact. Phys. Rev. Lett. 31, 22 (1973).

[13]   G.A. BAKER, Jr, S. KRINSKY : Renormalization group structure for translationally invariant ferromagnets. Journ. math. Phys. 18, 590 (1977).

The first rigorous mathematical work was done in the paper by Bleher and Sinai [5], on the case of a Gaussian fixedpoint (with $2^{\frac{1}{2}} < c < 2$).

The situation at that point was then clarified and reviewed in the following papers.

[14]   G. GALLAVOTTI, H. KNOPS : The Hierarchical Model and the
                renormalization group. Nuovo Cimento 5,
                341-368 (1975).

[15]   H. van BEYEREN, G. GALLAVOTTI, H. KNOPS : Conservation laws
                in the Hierarchical Model. Physica 78, 541
                (1974).

3. The Existence of a Non-Trivial Fixed Point

According to the Method M 1 we are looking for fixed points of the transformation $f \rightarrow \mathcal{N}_P^{(\beta)}(f)$. The word "non-trivial" of the title of this section indicates that we are interested particularly in fixed points which are neither Gaussians, or constant functions, nor δ-functions. It is a priori not at all clear that there should be any non-trivial fixed points. As an example, the transformation defined by Eq.(2.7) does not have non-trivial fixed points : in a sense this is exactly the content of the central limit theorem.

Let $f^{(\beta)}$ be a fixed point of the equation $f^{(\beta)} = \mathcal{N}_P^{(\beta)}(f^{(\beta)})$. Then $F^{(\beta)}(z) = \exp(-\beta \, f^{(\beta)}(z))$ satisfies the equation

$$F^{(\beta)}(z) = 2c^{-\frac{1}{2}} e^{\beta z^2/2} \int du \, F^{(\beta)}(z \, c^{-\frac{1}{2}} + u) F^{(\beta)}(z \, c^{-\frac{1}{2}} - u) \; .$$

$$(3.1)$$

We shall look therefore for fixed points of Eq.(3.1). We have arbitrarily fixed the constant in (3.1) to be equal to $\pi^{-\frac{1}{2}}$ which does not matter in the discussion of the equation because both sides are homogeneous of different degree.

We next eliminate the β dependence. If $F^{(\beta)}$ is a fixed point of Eq. (3.1), set (for $c \neq 2$)

$$\varphi(z) = S_\beta^{-1}(F^{(\beta)})(z)$$

$$= (4\pi(2-c)/c\beta)^{\frac{1}{2}} \exp(z^2/2) \, F^{(\beta)}(((2-c)/c\beta)^{\frac{1}{2}}z) \; .$$

$$(3.3)$$

Then φ is a solution of the following equation

$$0 = \pi^{-\frac{1}{2}} \int_{-\infty}^{+\infty} du \, e^{-u^2} \varphi(z \, c^{-\frac{1}{2}} + u) \varphi(z \, c^{-\frac{1}{2}} - u) - \varphi(z)$$

$$= \mathcal{N}(c,\varphi)(z) - \varphi(z) \qquad\qquad (3.2)$$

$$= \mathcal{F}(c, \varphi)(z) \ .$$

Conversely, since $\mathcal{S}_\beta^{-1}$ has an inverse, denoted $\mathcal{S}_\beta$, every solution of (3.2) yields a fixed point of (3.1) and hence of $\mathcal{N}_P^{(\beta)}$. But also the map $f \rightarrow \mathcal{N}_P^{(\beta)}(f)$ can be discussed in terms of $\mathcal{N}(c, \varphi)$, because

$$\exp(-\beta \ \mathcal{N}_P^{(\beta)}(f)) \quad = \quad \mathcal{S}_\beta \mathcal{N}(c, \ \mathcal{S}_\beta^{-1}(e^{-\beta f})) \ . \qquad (3.4)$$

Henceforth, we shall consider $\mathcal{N}(.,.)$ and $\mathcal{F}(.,.)$ as our main objects.

Before proceeding further let us remark two trivial but important properties.

1) If f is measurable, $f \geqslant 0$, $f \neq 0$ then $\mathcal{N}(c,f) > 0$, or $\mathcal{N}(c,f) = \infty$.

2) If $f(x) = f(-x)$ then $\mathcal{N}(c,f)(x) = \mathcal{N}(c,f)(-x)$, in other words, the space of even functions is invariant under $\mathcal{N}(c,.)$.

In order to proceed to Method M 1 we should now look for non-trivial solutions of $\varphi = \mathcal{N}(c,\varphi)$. This is not so easy, because $\mathcal{N}$ is non-linear, but we shall construct non-trivial solutions as bifurcations from trivial solutions at least in perturbation theory. So let us first discuss the "trivial" solutions.

Solution 1 : Gaussian case.

For all $c > 0$, $\varphi = 1$ is a solution of Eq. (3.2) . We call it Gaussian, because $P^{(\beta)}$ will be a Gaussian according to Eq. (3.3). The tangent map $\mathcal{D}_2 \mathcal{F}^*$ to $\mathcal{F}$ at $\varphi = 1$ will play a crucial role in the sequel and so we shall discuss it in detail.

In general, let φ_0 be given and let $\mathcal{D}_2 \mathcal{F}(c,\varphi_0)$ be the tangent map to $\mathcal{F}(c, \varphi)$ at $\varphi = \varphi_0$. Substituting $\varphi + \delta\varphi$ in (Eq.3.2) one sees that

*$\mathcal{D}_2$ denotes the "derivative" with respect to the second argument.

$$\mathcal{D}_2 \mathcal{F}(c, \varphi)(\delta\varphi)(z)$$

$$= \pi^{-\frac{1}{2}} \int du\ e^{-u^2} \varphi(z/\sqrt{c} - u)\delta\varphi(z/\sqrt{c} + u)$$
$$+ \pi^{-\frac{1}{2}} \int du\ e^{-u^2} \varphi(z/\sqrt{c} + u)\delta\varphi(z/\sqrt{c} - u) - \delta\varphi(z)$$

$$= 2\pi^{-\frac{1}{2}} \int du\ e^{-u^2} \varphi(z/\sqrt{c} + u)\delta\varphi(z/\sqrt{c} - u) - \delta\varphi(z). \qquad (3.5)$$

Coming back to our Solution 1 we have

$$\mathcal{D}_2 \mathcal{F}(c, 1)(g)(z) = 2\pi^{-\frac{1}{2}} \int du\ e^{-(z/\sqrt{c} - u)^2} g(u) - g(z) \qquad (3.6)$$

$$= (\ \mathcal{D}_2 \mathcal{N}(c, 1)(g))(z) - g(z)\ .$$

Let us consider the first term in (3.6) which is also equal to $\mathcal{D}_2 \mathcal{N}(c, 1)(g)$.

THEOREM 3.1 . The operator $\mathcal{D}_2 \mathcal{N}(c, 1)$ is selfadjoint on $L_2(\mathbb{R}, \exp(-\gamma x^2)dx)$, $\gamma = (1 - 1/c)$, and has eigenvectors $H_n(\gamma^{\frac{1}{2}}x)$ and eigenvalues $2/c^{n/2}$, where H_n is the n-th Hermite polynomial.

The Hermite polynomials are defined by

$$H_n(x) = (-1)^n e^{x^2} \partial_x^n e^{-x^2}\ , \quad n = 0, 1, \dots ; \qquad (3.7)$$

$$H_0(x) = 1, \quad H_1(x) = 2x\ , \quad H_2(x) = 4x^2 - 2\ , \quad \dots \ .$$

The corresponding orthonormal functions in $L_{2,\gamma} = L_2(\mathbb{R}, e^{-\gamma x^2} (\gamma/\pi)^{\frac{1}{2}}dx)$ are

$$h_n(c, x) = H_n(\gamma^{\frac{1}{2}}x) / (2^n n!)^{\frac{1}{2}} \qquad (3.8)$$

The proof of Theorem 3.1 is obtained by the substitution $c = 1/a^2$ in the following

LEMMA 3.2 . For $0 < a < 1$ the operator $f \rightarrow \pi^{-\frac{1}{2}} \int \exp-(az-u)^2 f(u)du$ is selfadjoint on $L_{2,(1-a^2)}$ and has spectrum a^n with eigenvectors $H_n((1-a^2)^{\frac{1}{2}}z)$, $n = 0, 1,...$ (cf Section 10 for a proof).

There is a second solution of Eq. (3.2).

Solution 2 : Delta function.

For all $c > 0$, $\varphi(z) = (4\pi/c)^{\frac{1}{2}}\delta(z)$ is a solution of Eq.(3.2). In order to discuss the spectrum of the tangent map, we prefer to go to Fourier transforms because there the function spaces seem better adapted. The Fourier transformed version of Eq. (3.2) is easily seen to be

$$\widetilde{\varphi}(p) = (c/2\pi)^{\frac{1}{2}} \int dq \ e^{-q^2} \widetilde{\varphi}(pc^{\frac{1}{2}}/2 - q) \ \widetilde{\varphi}(pc^{\frac{1}{2}}/2 + q)$$

$$= \widetilde{\mathcal{N}} (c,\widetilde{\varphi})(p) ,$$

so that the constant solution $\widetilde{\varphi}(p) = (2/c)^{\frac{1}{2}}$ corresponds to the δ function described above. Also $\widetilde{\mathcal{N}}(c,\widetilde{\varphi}) = (c/2)^{\frac{1}{2}} \mathcal{N}(4/c,\widetilde{\varphi})$. Using again Lemma 3.2 we get

LEMMA 3.3 . The operator $D_2\widetilde{\mathcal{N}}(c,(2/c)^{\frac{1}{2}})$ is selfadjoint on $L_{2,(1-c/4)}$ and has eigenvectors $H_n((1-c/4)^{\frac{1}{2}}p)$ and eigenvalues $2 \cdot (c/4)^{n/2}$, $n = 0, 1, 2, ...$.

The two solutions found above will be used often later. But, as we have already said before, we are more interested in non-trivial solutions of Eq. (3.2) which occur for $c < 2^{\frac{1}{2}}$. Such solutions (maybe not all of them) can be found essentially without guesswork by using the theory of bifurcation from simple eigenvalues. We restate this theory for the convenience of the reader, in the framework of Hilbert spaces. However, it will turn out that this framework is much too narrow for the problem at hand and that new methods will be needed to show the existence of a non-trivial solution outside of perturbation theory which is called ε -expansion in the physics literature. On the

other hand, the theory will allow for a compact definition of the
ε-expansion and it will show that the ε-expansion is the perturbation
theory of a bifurcation.

Let $\mathcal{F}: V \times \mathcal{K} \rightarrow \mathcal{K}$ be a continuous map from a neighborhhod $V \subset \mathbb{R}$
of c_0 and a Hilbert space $\mathcal{K}$ into $\mathcal{K}$, and assume $\mathcal{D}_2\mathcal{F}$, and $\mathcal{D}_1\mathcal{D}_2\mathcal{F}$ exist
and are continuous. Suppose further that the following three condi-
tions are met

1) $\mathcal{F}(c, \varphi_0) = 0$ for all $c \in V$.

   (In our case, $\varphi_0 = 1$) .

2) $\mathcal{D}_2\mathcal{F}(c_0, \varphi_0)$ has simple isolated eigenvalue 0 with eigenvector v.

3) $\mathcal{D}_1\mathcal{D}_2\mathcal{F}(c_0, \varphi_0)v$ is not orthogonal to v.

THEOREM 3.4. Under the above conditions there are two continuous
functions of α in a neighborhood U of $0, \mu : U \rightarrow \mathbb{R}$, and
$\psi : U \rightarrow v^\perp \subset \mathcal{K}$ (the subspace of $\mathcal{K}$ orthogonal to v) such that

$$\mathcal{F}(c_0 + \mu(\alpha), \varphi_0 + \alpha v + \alpha \psi(\alpha)) = 0 , \qquad (3.9)$$

and $\mu(0) = \psi(0) = 0$.

In other words, there is a second solution to $\mathcal{F}(c, \varphi) = 0$ in addition to the trivial one ($\varphi = \varphi_0$) and it bifurcates away from the
trivial one in the direction of the eigenvector v whose eigenvalue is
zero at $c = c_0$.

We give here the proof up to the point where it provides us with
an algorithm which tells how to calculate the solution, while for the
missing step (the implicit function theorem) the reader is referred
to the mathematics literature. One defines a function

$$f(\alpha, \mu, \psi) = \begin{cases} \alpha^{-1} \, \mathcal{F}(c_0 + \mu, \, \varphi_0 + \alpha v + \alpha \psi) & \text{if } \alpha \neq 0 \\ \mathcal{D}_2\mathcal{F}(c_0 + \mu, \varphi_0)(v + \psi) & \text{if } \alpha = 0 \end{cases} . \qquad (3.10)$$

By construction $f(0, 0, 0) = 0$ and the Fréchet derivative of the map $(\mu, \psi) \rightarrow f(0, \mu, \psi)$ at $(\mu, \psi) = (0, 0)$ is the linear map

$$(\mu^*, \psi^*) \quad \rightarrow \mu^* \, D_1 D_2 \mathcal{F}(c_o, \varphi_o) v + D_2 \mathcal{F}(c_o, \varphi_o) \psi^* \ . \tag{3.11}$$

The hypotheses 2) and 3) imply that this map is an isomorphism, that is it has an inverse, and hence the implicit function theorem (on Hilbert spaces) can be applied to the equation

$$f(a, \mu(a), \psi(a)) = 0 \ . \tag{3.12}$$

This yields a solution which is necessarily non-trivial (i.e. $\neq \varphi_o$) and one can also show its uniqueness.

We proceed now to do perturbation theory. For this it is of course necessary to assume that $f(a, \mu, \psi)$ has derivatives of arbitrary order. We shall see later that this is indeed the case for our particular function $\mathcal{N}(c, \varphi)$ defined in Eq.(3.2), so that the present discussion will apply. Then one can solve (3.12) in perturbation theory, and this can be done by iteration. The equations are however somewhat tedious to write down in the general case, but we shall do it now for a particularly important case in our example.

It is useful to write the operator $\mathcal{N}_c = \mathcal{N}(c, .)$ in a Hermite basis. Here, and throughout the Lecture Notes, it is sometimes convenient to view the quadratic map $\varphi \rightarrow \mathcal{N}_c(\varphi)$ as the restriction to the diagonal (i.e. to equal arguments) of a bilinear map denoted again $\mathcal{N}_c$ and given by

$$\mathcal{N}_c(f, g)(z) \quad = \quad \pi^{-\frac{1}{2}} \int_{-\infty}^{+\infty} du \, e^{-u^2} f(zc^{-\frac{1}{2}} - u) g(zc^{-\frac{1}{2}} + u). \tag{3.13}$$

One finds by integration by parts, using the orthogonality of the $h_n(c, x)$ on $L_{2, \gamma}$ (cf.(3.8)) ,

$$(h_k, \mathcal{N}_c(h_n, h_{n'}))_{2,\gamma}$$

$$
= \begin{cases}
\dfrac{1}{c^{k/2}} \dfrac{(\frac{2}{c} - 1)^{(n+n'-k)/2}}{((n+n'-k)/2)!} \left(\dfrac{k}{\frac{n-n'+k}{2}} \right) \left(\dfrac{n!\, n'!}{k!} \right)^{\frac{1}{2}}, & \text{(3.14)} \\[1em]
\qquad \text{if } |n - n'| \leqslant k,\ n + n' \geqslant k \text{ and } n + n' + k \text{ even,} \\[0.5em]
0 \qquad \text{otherwise,}
\end{cases}
$$

where $(\ ,\ )_{2,\gamma}$ is the scalar product on $L_{2,\gamma}$.

It is convenient to set

$$\rho_j(c,x) = h_j(c,x) \,/\, \left(j!\ (2/c - 1)^j \right)^{\frac{1}{2}}, \tag{3.15}$$

and to write

$$\varphi(c(\alpha), z) = \sum a_j(\alpha) \rho_j(c(\alpha), z) . \tag{3.16}$$

Now the equation $\varphi = \mathcal{N}_c(\varphi)$ becomes, using Eqs. (3.14), (3.15)

$$a_k(\alpha) = \sum_{\substack{|n-n'| \leqslant k \\ n+n' \geqslant k \\ n+n'+k \text{ even}}} a_n(\alpha)\, a_{n'}(\alpha) \left(\dfrac{k}{\frac{n-n'+k}{2}} \right) \left(c^{k/2}(\alpha) \left(\dfrac{n+n'-k}{2} \right)! \right)^{-1} \tag{3.17}$$

According to hypothesis 2, the points of interest are those values of c for which $\mathcal{D}_2 \mathcal{F}(c, \varphi_0 = 1)$ has an eigenvalue equal to zero. From a physical point of view, the most interesting case is the case when $2/c^2 - 1$, the eigenvalue corresponding to H_4 is zero. The case $2/c^0 - 1 = 0$ does not occur and the case $2/c^1 - 1 = 0$ corresponds to a quadratic change of Hamiltonian which is usually absorbed into a change of the variance of the mean spin. In the case $c_o = 2^{\frac{1}{2}}$ we

shall parametrize the neighborhood of this point by $c = 2^{\frac{1}{2}(1-\varepsilon)}$(other

parametrizations will be occasionally taken to simplify the notation).

This choice of parametrization is motivated by the following heuristic

consideration. The "critical dimension" is usually computed according

to the so-called Ginzburg argument which we repeat for convenience.

Note that this is based on dimensional analysis which does not claim

any rigour otherwise. Consider an interaction potential of the form

$r^{-d-\sigma}$ where r is the distance, d is the dimension of the space and

σ controls the range. The case $\sigma > 2$ is called "short-range", but the

Hierarchical Model is a model with "long-range". Indeed, d = 1 and we

have seen in Section 2 that the potential is about $r^{\log_2(c/4)}$, i.e.

$\sigma = -\log_2(c/4)-1 = 1 - \log_2 c$. Since we allow c in the interval

$1 < c < 2$, we have $0 < \sigma < 1$. The so called Ginzburg criterion sti-

pulates that non-trivial critical indices (and hence nontrivial fixed

points) for a thermodynamic function with single spin density

$\exp(-m\ s^2 - u_p s^p)$ can only occur if $d/\sigma < p/(p-2)$. Now the "bifurcation

direction" from $c = 2^{\frac{1}{2}}$ is the polynomial H_4 and we have p = 4 in lowest

order perturbation theory. If now $c = 2^{\frac{1}{2}(1-\varepsilon)}$, then $\sigma = \frac{1}{2} + \varepsilon/2$ and

the condition reads $d < (4/2) \cdot (\frac{1}{2} + \varepsilon/2)$, i.e. $d < 1 + \varepsilon$. The "criti-

cal value" for d is thus $1 + \varepsilon$ (called critical dimension), while the

actual dimension (which is 1) is by ε below the critical dimension

hence we are expanding a solution which is by ε dimensions below the

critical dimension, in analogy with the short range case, $(\sigma = 2)$,

where the critical dimension is 4 and the actual dimension is $4 - \varepsilon$.

If we set now $c_o = 2^{\frac{1}{2}}$, $\mu(o) = 0$, $c(\alpha) = c_o + \mu(\alpha) = 2^{\frac{1}{2}(1-\varepsilon(\alpha))}$,

$a_j(0) = \delta_{jo}$, and $a_4(\alpha) = \alpha$, $a_j(\alpha) = 0 + O(\alpha^2)$, $j \neq 4$, $j \neq 0$ then this

is nothing but the ansatz (3.10) with $v = \rho_4$. Eq. (3.17) can now be

solved by recursion. Substituting the values obtained so far to order n in α into the RHS yields them to order $n + 1$, except for the term $k = 4$, which is used to determine c to the next order. The result is

$$a_o(\alpha) = 1 - \alpha^2/24 + \mathcal{O}(\alpha^3),$$

$$a_2(\alpha) = -\alpha^2(3(2-2^{\frac{1}{2}}))^{-1} + \mathcal{O}(\alpha^3),$$

$$a_6(\alpha) = -\alpha^2 \cdot 10(2^{\frac{1}{2}}-1)^{-1} + \mathcal{O}(\alpha^3),$$

$$a_8(\alpha) = +\alpha^2 \cdot 35 + \mathcal{O}(\alpha^3),$$

$$a_{2k}(\alpha) = \mathcal{O}(\alpha^3), \quad k = 5,6,7,\ldots,$$

$$\varepsilon(\alpha) = -\alpha \cdot 3(2 \log 2)^{-1} - \alpha^2((17+18 \ 2^{\frac{1}{2}}) \ (3 \log 2)^{-1}) + \mathcal{O}(\alpha^3).$$

It remains now to solve for ε (by inversion of the power series), and to express each ρ_j as a formal power series in ε and H_{2k}, $k=1,\ldots j$. This is done easily, using the definition of Hermite polynomials. One notes here that only a finite number of $a_j(\alpha(\varepsilon))\rho_j$, contribute to a term $\varepsilon^n H_{2k}$.

One gets with $H_n = H_n((1 - 2^{-\frac{1}{2}})^{\frac{1}{2}}x)$,

$$\varphi_\varepsilon(x) = 1 - \varepsilon \log 2/(144(2^{\frac{1}{2}}-1)^2) \ H_4$$

$$+ \ \varepsilon^2 \ \{ \ - \frac{(\log 2)^2}{54} \ + H_2(\ - \frac{(\log 2)^2}{2^{\frac{1}{2}} \ 27(2^{\frac{1}{2}} -1)^2} \ + \ \frac{(\log 2)^2}{24(2^{\frac{1}{2}}-1)^3} \ )$$

$$+ \ H_4(\ - \frac{(\log 2)^2}{(2^{\frac{1}{2}}-1)^2} \ \frac{17 + 18 \cdot 2^{\frac{1}{2}}}{972} \ + \ \frac{(2^{\frac{1}{2}}+1)(\log 2)^2}{144 \ (2^{\frac{1}{2}}-1)^3} \ )$$

$$+ \ H_6 \ \frac{(\log 2)^2}{(2^{\frac{1}{2}}-1)^4 \ 1296} \ + \ H_8 \ \frac{(\log 2)^2}{(2^{\frac{1}{2}}-1)^4 \ 41472} \ \} + \mathcal{O}(\varepsilon^3) \ .$$

We shall write this as

$$\varphi_\varepsilon(x) = 1 - \varepsilon \ \$' \ H_4((1 - 2^{-\frac{1}{2}})^{\frac{1}{2}}x)+\mathcal{O}(\varepsilon^2), \text{ with } \$' = \log 2/(144(2^{\frac{1}{2}}-1)^2)$$

$$= 1 - \varepsilon \ \$ \ \{x^4 - 3x^2/(1 - 2^{-\frac{1}{2}}) + 3/(2(2^{\frac{1}{2}} - 1)^2) \ \} + \mathcal{O}(\varepsilon^2) \text{ with}$$

$$\$ = 8(2^{\frac{1}{2}} -1)^2 \$' \ .$$

The constants $\$$, $\$'$ will always designate these particular values.

In Section 9 we present the output of a computer program which produces the ε-expansion along similar lines up to order 34, and which gives in particular the expansion of the second eigenvalue of the tangent map along the new branch.

So far, we have only discussed the existence of a perturbation series for the function φ, which is a non-trivial solution of $\varphi = \mathcal{N}_c'(\varphi)$. We shall concentrate in the sequel on the case $c(\varepsilon) = 2^{\frac{1}{2}(1-\varepsilon)}$, and we shall write $\mathcal{N}_\varepsilon$ for $\mathcal{N}_{c(\varepsilon)}$, $\mathcal{DN}_\varepsilon$ for $\mathcal{D}_2\mathcal{N}(c(\varepsilon), 1)$, $\mathcal{DN}_{\varphi,\varepsilon}$ for $\mathcal{D}_2\mathcal{N}(c(\varepsilon),\varphi)$, $\mathcal{N}_\varepsilon'(f,g)$ for the bilinear form $\mathcal{N}_{c(\varepsilon)}'(f,g)$ and φ_ε for the non-trivial solution of the equation $\varphi_\varepsilon = \mathcal{N}_\varepsilon(\varphi_\varepsilon)$, if it exists. While expansions give us a computational tool to work with, they do not show the existence of the object in question. A particularly simple example of a phenomenon of this kind is given by the series $\Sigma \varepsilon^n n!$ which defines no object in the class of functions analytic at $\varepsilon = 0$. We cannot be content with a power series alone, because we really need to know whether a (non-trivial) fixed point exists as a thermodynamic object. For example, the perturbation theory for the phenomena at the critical temperature could exist while the model in question would not have phase transitions.

We therefore want to <u>prove existence</u> of a solution φ_ε . It now turns out that this is a much harder question than to show the existence of a perturbation theory. Namely on all of the "reasonable" spaces either $\mathcal{N}$ is not continuous or $\mathcal{D}_2\mathcal{N}$ does not have discrete spectrum. So while we have seen that (3.12) most elegantly leads to a solution of the problem in perturbation theory, not even the most advanced versions of the implicit function theorem seem to be sufficient for our problem. The reason for this is on one hand the unboundedness of $\mathcal{N}$, which can be seen explicitly from the factor $\binom{k}{(n-n'+k)/2}$

in (3.17). On the other hand, whenever $\mathcal{N}$ is bounded, $\mathcal{D}_2\mathcal{N}$ does not seem to have eigenvalues. Put in another way, neighborhoods are too big in function spaces. One could also say that the polynomial approximation $\varphi(\alpha) \sim 1 + \alpha \, \rho_4$ does not push us sufficiently into the direction of the bifurcating branch for the implicit function thm. to apply.

We shall therefore try a better initial approximation and this will be sufficient after some hard labour. Instead of writing

$$\varphi_\varepsilon(x) = 1 - \varepsilon \, \$' \, H_4(x) + \mathcal{O}(\varepsilon^2), \tag{3.18}$$

we make the ansatz

$$\varphi_\varepsilon(x) = e^{-\varepsilon\$ \, x^4} \, P_N(\varepsilon,x) + \text{remainder} \tag{3.19}$$

$$= f_N(\varepsilon,x) + \text{remainder} ,$$

where $P_N(\varepsilon,x)$ is that polynomial in ε,x such that $f_N(\varepsilon,x)$ coincides with φ_ε up to order N in ε. The existence (and uniqueness) of such a polynomial follows from our previous considerations on the perturbation theory for φ_ε and from the invertibility of the exponential function.

LEMMA 3.5 . The coefficient of ε^k in $P_N(\varepsilon,x)$ is a polynomial of degree $\leqslant 2k$ in x for k = 0, ..., N. Furthermore

$$\mathcal{N}_\varepsilon(f_N(\varepsilon,.))(x) - f_N(\varepsilon,x) = \exp(-2 \, \varepsilon\$x^4/3) \, g_\varepsilon(x) , \tag{3.20}$$

with $|g_\varepsilon(x)| \leqslant \mathcal{O}(\varepsilon^{N/2})$. \tag{3.21}

Deferring the proof to Section 8, we shall now state the main estimate which leads to the existence of $\varphi_\varepsilon(x)$.

Write $\varphi_\varepsilon(x) = f_N(\varepsilon,x) + R_\varepsilon(x)$. We shall look for a small R_ε in L_∞ such that

$$\mathcal{N}_\varepsilon(\varphi_\varepsilon) = \varphi_\varepsilon \quad . \tag{3.22}$$

Using the quadratic nature of $\mathcal{N}_\varepsilon$, and the definition of $\mathcal{D}_2\mathcal{N}(c(\varepsilon),\psi)\psi'$ $= 2\,\mathcal{N}_\varepsilon(\psi,\psi')$ cf. (3.13), we set $\mathcal{A}_{\psi,\varepsilon} = \mathcal{A}_\psi = \mathcal{D}_2\mathcal{N}(c(\varepsilon),\psi)$. We get with $f_N = f_N(\varepsilon,.)$,

$$f_N + R_\varepsilon = \mathcal{N}_\varepsilon(f_N) + \mathcal{N}_\varepsilon(R_\varepsilon) + \mathcal{A}_{f_N,\varepsilon}R_\varepsilon \tag{3.23}$$

and solving for the part linear in R_ε

$$R_\varepsilon = (\mathcal{A}_{f_N,\varepsilon} - 1)^{-1}\{ f_N - \mathcal{N}_\varepsilon(f_N) - \mathcal{N}_\varepsilon(R_\varepsilon) \} \quad . \tag{3.24}$$

Our first main estimate is the

THEOREM 3.6. For all $N \geqslant 0$, there is an $\varepsilon_0(N) > 0$ such that for $0 < \varepsilon < \varepsilon_0(N)$ the operator $(\mathcal{A}_{f_N,\varepsilon} - 1)^{-1}$ is bounded from L_∞ to L_∞ with norm less than $C_N\,\varepsilon^{-12}$. (In Part II a stronger theorem is proved.)

Therefore, if N is such that $|f_N - \mathcal{N}_\varepsilon(f_N)|(x) \leqslant \mathcal{O}(\varepsilon^{25})$, (i.e $N \geqslant 50$), the map

$$R \rightarrow (\mathcal{A}_{f_N,\varepsilon} - 1)^{-1}(f_N - \mathcal{N}_\varepsilon(f_N) - \mathcal{N}_\varepsilon(R))$$

is a contraction of the ball of radius $\mathcal{O}(\varepsilon^{25})$ in L_∞ and (3.22) possesses thus a unique fixed point. We have thus shown the existence of φ_ε. In fact, $\varphi_\varepsilon = f_N(\varepsilon,.) + \mathcal{O}(\varepsilon^{(N-1)/2})$ in L_∞. Note that this is only existence for $\varepsilon > 0$, so we do not have a bifurcation into two solutions as would be implied by perturbation theory alone.

We sketch here the main ideas of the proof of Theorem 3.6 , because it should have some interest of its own. First we observe that

$\mathcal{A}_{f_N, \varepsilon}$ has two distinct important features.

i) $\mathcal{A}_{f_N, \varepsilon}$ has an integral kernel which decays like

$\exp(-(zc^{-\frac{1}{2}} - u)^2 - \$\varepsilon(2zc^{-\frac{1}{2}} - u)^4)$ and from this we get the bound

$$\| \mathcal{A}_{f_N, \varepsilon} g \|_\infty \leq K_N \|g\|_{p, \gamma} \quad ,$$

if $p \geq 2\gamma\varepsilon^{-1}$ (cf. Lemma 10.5).

ii) The function f_N is near 1 in $L_{4, \gamma}$ and hence (cf. Lemma 10.3), the operator $\mathcal{A}_{f_N, \varepsilon}$ is near to $\mathcal{A}_{1, \varepsilon}$. But $\mathcal{A}_{1, \varepsilon}$ has the property of regularizing functions ("hypercontractivity") in the sense that

$$\| \mathcal{A}_{1, \varepsilon} g \|_{p+1, \gamma} \leq 2\|g\|_{pc^{-1}+1, \gamma} \quad ,$$

(cf. Lemma 10.6) .

The proof of Theorem 3.6 is now basically as follows. Let P_j be the spectral projection corresponding to one of the eigenvalues λ_j of $\mathcal{A}_{f_N, \varepsilon}$. By perturbation theory we have $|\lambda_j - 1|^{-1} \leq \mathcal{O}(\varepsilon^{-1})$. Then

$$\| (\mathcal{A}_{f_N, \varepsilon} - 1)^{-1} P_j g \|_\infty$$

$$= \| (\lambda_j - 1)^{-1} P_j g \|_\infty$$

$$= | (\lambda_j - 1)^{-1} \lambda_j^{-n(\varepsilon)} | \; \| \mathcal{A}_{f_N, \varepsilon}^{n(\varepsilon)} P_j g \|_\infty$$

$$\leq | (\lambda_j - 1)^{-1} \lambda_j^{-n(\varepsilon)} | \; \mathcal{O}(1) \| \mathcal{A}_{f_N, \varepsilon}^{n(\varepsilon)-1} P_j g \|_{2\gamma\varepsilon^{-1} + 1, \gamma}$$

$$\leq | (\lambda_j - 1)^{-1} \lambda_j^{-n(\varepsilon)} | \; \mathcal{O}(1) \; 4^{n(\varepsilon)-1} \| P_j g \|_{2\gamma\varepsilon^{-1} c^{-n(\varepsilon)+1} + 1, \gamma}$$

by a repeated application of (ii) . For $n(\varepsilon) \sim \log_c \varepsilon^{-1}$, we have $2\gamma\varepsilon^{-1} c^{-n(\varepsilon)+1} + 1 \leq 2$ so that we have

$$\| (A_{f_N, \varepsilon} - 1)^{-1} P_j g \|_\infty$$

$$\leqslant \quad | (\lambda_j - 1)^{-1} \lambda_j^{-n(\varepsilon)} | \; \mathcal{O}(1) \; 4^{n(\varepsilon)-1} \| P_j g \|_{2,\gamma}$$

$$\leqslant \quad \mathcal{O}(\varepsilon^{-k}) \| P_j g \|_{2,\gamma}$$

$$\leqslant \quad \mathcal{O}(\varepsilon^{-k}) \| g \|_{2,\gamma}$$

$$\leqslant \quad \mathcal{O}(\varepsilon^{-k}) \| g \|_\infty \quad .$$

Repeating this argument for the eigenvalues near 1 and on the spectrum near 0, one gets the result.

It will be necessary and useful to prove detailed statements about the function R_ε .

THEOREM 3.7 . The function φ_ε has the following properties (for $\varepsilon > 0$ sufficiently small)

1) $| \varphi_\varepsilon(x) | \leqslant \mathcal{O}(\exp(-\varepsilon \$ \, x^4 / 2))$ as $\quad x \to \infty$.

2) $\varphi_\varepsilon(x)$ is an entire function in x, and in the class $\mathscr{S}^{1/4 - \mathcal{O}(\varepsilon)}_{3/4 + \mathcal{O}(\varepsilon)}$ of Gelfand- Shilov [19] .

3) As a function of $\varepsilon \geqslant 0$, φ_ε is a real analytic function for $\varepsilon > 0$ which is infinitely differentiable at $\varepsilon = 0$ (setting $\varphi_0 = 1$) .

(See Theorem 11.1 for a detailed definition of the function spaces). The proofs of these statements are in two main steps. First one repeats a proof of Theorem 3.6 in a space of functions which decrease like $\exp(-\varepsilon \$ \, x^4 / 2)$ and which are once continuously differentiable in x. Formulas very similar to (3.24) appear at this stage. The higher successive derivatives in x are obtained by differentiating both sides of (3.22) and integrating by parts on the RHS. Then the LHS (say an n'th derivative) is expressed in terms of lower derivatives only. The analyticity properties follow. The derivatives in ε are obtained in a similar way, but one has to invert in addition $(\mathcal{A}_{1,\varepsilon} - 1)$.

We have thus a rather detailed knowledge of the critical spin distribution φ_ε . It is the point of view taken in these Lecture Notes that this knowledge __alone__ is sufficient to describe the physics of the Hierarchical Model i.e. its __critical indices__. These are then the results of Method 1 . It is only in Method 2 that another fixed point (the Solution 2) will play a role, and that the "crossover" (the flow under $\mathcal{N}$ from the fixed point φ_ε to the fixed point $(4\pi/c)^{\frac{1}{2}}\delta$) has to be studied. This will yield the proof of the existence of the thermo-dynamic limit and the presence of a phase transition.

All the above considerations have a straightforward generaliza-tion to points $c = 2^{1/j}$, $j = 2,3,\ldots$. The bifurcating solution is then of the form $\exp(-\varepsilon \$_j x^{2j})\cdot$polynomial , and again a fixed point theorem applies. We do not expect this to happen at the points $c = 2^{\frac{1}{(2j+1)/2}}$, because there the natural ansatz is $\exp(\pm\varepsilon \, a \, x^{2j+1})$ which is unbounded for $x \rightarrow \pm\infty$. For the case c near $2^{1/j}$ all prece-ding results hold, so that one has the bifurcation picture

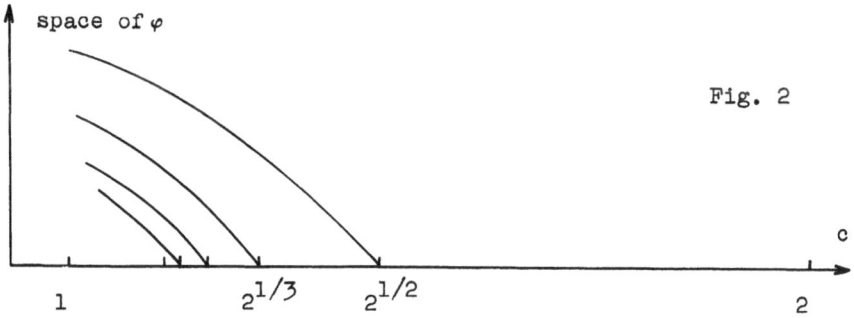

Fig. 2

where each branch is controlled near the branching point. These branches correspond to "critical", "tricritical", "tetracritical" ... behaviour, as $c = 2^{1/2}$, $2^{1/3}$, $2^{1/4}$ ... respectively. Bleher [20] has done numerical calculations and followed the critical branch almost to $c = 1$. He found no further bifurcations, and the following diagram shows how the function φ_ε behaves (as a function of c). The Hierarchical Model has no phase transition at $c = 1$ [10].

Fig.3. The critical spin distribution.

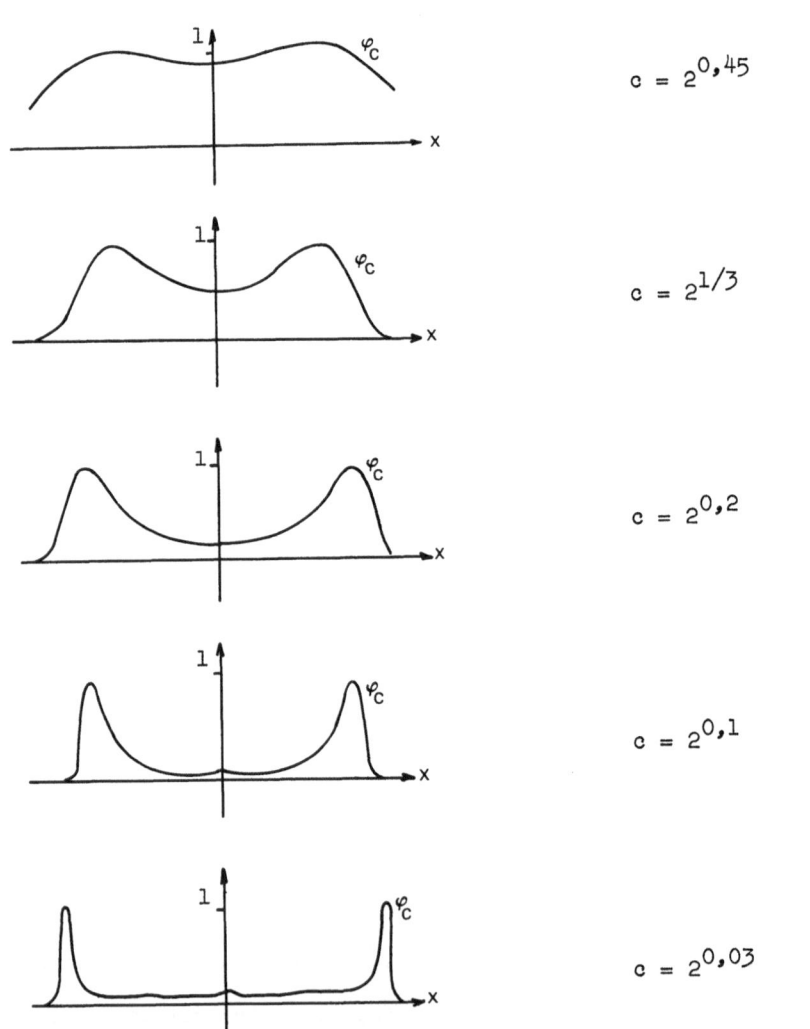

$$c = 2^{0,45}$$

$$c = 2^{1/3}$$

$$c = 2^{0,2}$$

$$c = 2^{0,1}$$

$$c = 2^{0,03}$$

We discuss now a property of φ_ε which is related to the probabilistic aspects of RG theory and which shows that φ_ε is "non-trivial" in a very strong sense. As we have said in Section 1, φ_ε can be viewed as the probability density for a suitably rescaled sum of dependent random variables.

PROPOSITION 3.8. The distribution $\exp(-\tfrac{1}{2} x^2) \, \varphi_\varepsilon(x)$ is not infinitely divisible. In particular this implies that this density is a new object and cannot be obtained as a limit of sums of independent random variables.

Proof : (Sketch). It follows from perturbation theory in ε that the truncated four point function is negative. However, it is positive for infinitely divisible distributions. Hence $\exp(-\tfrac{1}{2} x^2) \varphi_\varepsilon(x)$ is not infinitely divisible.

Remarks on Section 3 :

The breakthrough in the computation of a non-trivial fixed point was
the paper

[16]   P.M. BLEHER, Ja.G. SINAI : Critical Indices for Dyson's Asymp-
                totically Hierarchical Models, Commun. Math. Phys. $\underline{45}$
                347.(1975).

Theorem 3.1 is taken from the paper

[17]   M.G. CRANDALL, P.H. RABINOWITZ : Bifurcation from Simple Eigen-
                values. J. Funct. Anal. $\underline{8}$, 321 (1971).

The proof of Bleher and Sinaï used the "method of the separatrix".
Improving slightly on their method , we showed in

[18]   P. COLLET, J.-P. ECKMANN: The ε-Expansion for the Hierarchical
                Model. Commun. Math. Phys. $\underline{55}$, 67 (1967).

that φ_ε is a C^N function of $\varepsilon \geqslant 0$, for all N and ε sufficiently small,
so that the ε-expansion for the critical indices, and more knowledge
about φ_ε follows. The proof of the existence of φ_ε we give in these
Lectures Notes is new and has not appeared before. It relies on hyper-
contractive estimates, cf [36] known from constructive field theory.

The reference [19] is

[19]   I.M. GELFAND, G.E. SCHILOW : Verallgemeinerte Funktionen
                (Distributionen) Band II, Berlin, 1962, VEB Deutscher
                Verlag der Wissenschaften.

The results of the numerical calculations of Bleher can be found in

[20]   P.M. BLEHER : Critical indices for models with long range forces
            (Numerical Calculations). Preprint. Inst. of Applied
            Math., Acad. Sci. SSSR (1975).

The case $\sqrt{2} < c < 2$ has been discussed in great detail in

[21]   P.M. BLEHER : A second order phase transition in some ferroma-
            gnetic models. Trudy Mosc. Math. Obshestvo<u>33</u>, 155
            (1975).

The results on the critical indices have been summarized in

[22]   P.M. BLEHER, Ja.G. SINAI : Critical indices for systems with
            slowly decaying interaction. Zh.Eksp. Teor. Fiz. <u>67</u>
            391 (1974) [Sov. Phys. JETP. <u>40</u> , 195 (1975)].

Theorem 3.8. is a variant of an argument suggested by Nappi-Hegerfeldt
and given in

[23]   M. CASSANDRO, G. JONA-LASINIO : Asymptotic behaviour of the auto-
            covariance function and violation of strong mixing
            (Preprint).

[24]   G.C. HEGERFELDT : Prime field decompositions and infinitely
            divisible states on Borcher's tensor algebra. Commun.
            math. Phys. <u>45</u> , 137 (1975).

4. The Flow Around the Fixed Point

In this section , and the following, we <u>fix</u> ε to some (sufficiently small) positive value. Then the fixed point φ_{ε} is in L_{∞} and there
are standard methods to discuss the flow induced by the map

$$\psi \quad \to \mathcal{N}_{\varepsilon}^{\circ}(\psi + \varphi_{\varepsilon}) - \mathcal{N}_{\varepsilon}^{\circ}(\varphi_{\varepsilon}) =: \quad T(\psi) . \tag{4.1}$$

On Banach spaces, the flow around a fixed point can be almost
completely characterized by the tangent map at the fixed point. In
our case

$$\mathcal{D} T(\varphi_{\varepsilon})(\psi) = \mathcal{D} \mathcal{N}_{\varepsilon}^{\circ}(\varphi_{\varepsilon})(\psi) = 2 \mathcal{N} (\varphi_{\varepsilon}, \psi) . \tag{4.2}$$

cf. page 24 .

The spectrum of $2 \mathcal{N}_{\varepsilon}^{\circ}(\varphi_{\varepsilon}, \cdot)$ can be computed in perturbation theory ,
since, e.g. in $L_{2,\gamma}, \varphi_{\varepsilon} = 1 - \varepsilon \mathfrak{z}' H_4 (x(1-2^{-\frac{1}{2}})^{\frac{1}{2}}) + \mathcal{O}(\varepsilon^2)$ by the results of
the previous section. The result is the

<u>THEOREM 4.1</u> . <u>The spectrum of $2 \mathcal{N}_{\varepsilon}^{\circ}(\varphi_{\varepsilon}, \cdot)$ on L_{∞} consists, for suf-
ficiently small $\varepsilon > 0$ (depending on N) of eigenvalues</u>

$$2/2^{j/4} + \mathcal{O}(\varepsilon) \text{ for } j = 0, \ldots N, \ j \neq 4,$$

<u>and</u> $\qquad 1 - \varepsilon \log 2 + \mathcal{O}(\varepsilon^2) ,$

<u>and a remainder in the interval $[-2/2^{N/4}, 2/2^{N/4}]$ (the spectrum is
discrete), (cf Corollary 10.10).</u>

So, graphically, the spectrum takes the form

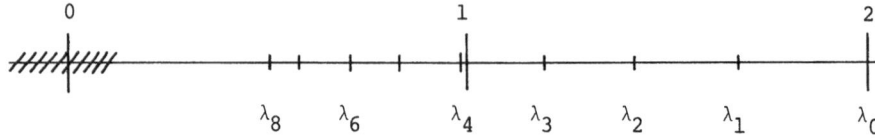

Fig. 4

Note that λ_4 is $1 - \epsilon \log 2 + O(\epsilon^2)$ for $\mathcal{D}\mathcal{N}_\epsilon(\psi_\epsilon)$ while the corresponding eigenvalue for $\mathcal{D}\mathcal{N}_\epsilon(1)$ is $2/c^2 \sim 1 + \epsilon \log 2$. Therefore the eigenvalue for $\mathcal{D}T$ is smaller than one ($\mathcal{D}T$ is a contraction in the "direction" associated to λ_4) while $\mathcal{D}\mathcal{N}_\epsilon(1)$ is an expansion in the analogous direction. So the bifurcating branch is in a sense more stable than the branch from which it bifurcates since the former has one more contractive direction than the latter.

Let now E_s and E_u be the spectral subspaces of $\mathcal{D}T$ corresponding to the eigenvalues less than one and greater than one, respectively, (s and u stand for stable and unstable resp.). The "flow" T can be stretched by an infinitely differentiable coordinate transformation S so that the following is true.

THEOREM 4.2.   There is a C^∞ diffeomorphism S on a (sufficiently small) ball $\mathcal{B} \subset L_\infty$ such that

$$S T S^{-1} = \mathcal{D}\mathcal{N}_\epsilon(\psi_\epsilon) + \mathcal{R}\,,$$

with a remainder $\mathcal{R}$ which is C^∞ and satisfies

$$\| \mathcal{R}(f) \|_\infty \leqslant \|f\|_\infty^{3/2}\,, \qquad\qquad (4.3)$$

and which satisfies $\mathcal{R} E_u \subset E_u$, $\mathcal{R} E_s \subset E_s$.

The norm condition (4.3) implies that $\mathcal{R}$ is small in norm as one gets near the fixed point.

We can visualize the statement of the theorem as follows. The map T has a stable and an unstable manifold $\mathcal{W}_s$ and $\mathcal{W}_u$ (resp.) corresponding to the eigenvalues of $\mathcal{D}T = \mathcal{D}\mathcal{N}_\epsilon(\varphi_\epsilon)$ which are smaller and bigger than one, respectively.

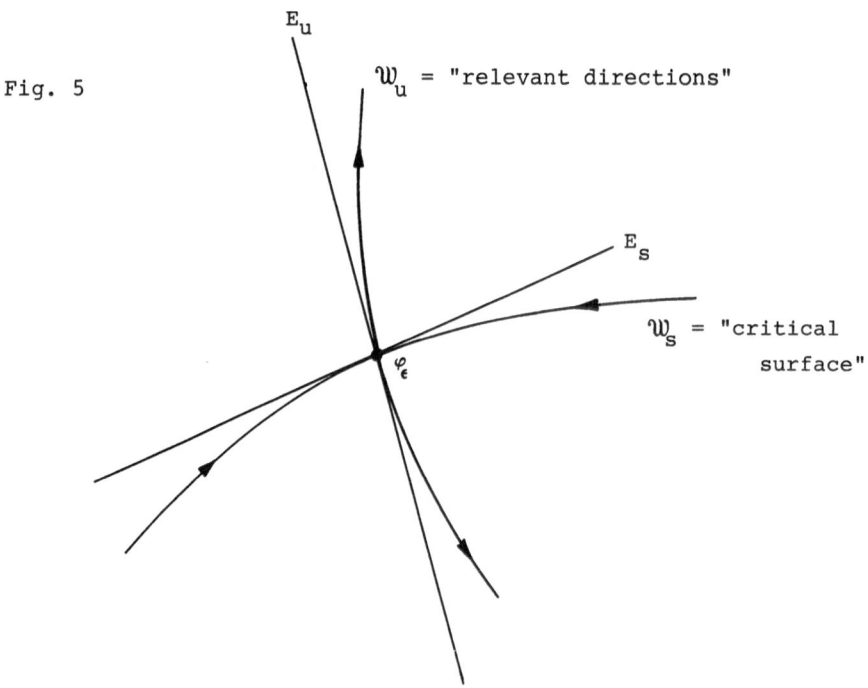

Fig. 5

$\mathcal{W}_u$ = "relevant directions"

E_u

E_s

$\mathcal{W}_s$ = "critical surface"

φ_ϵ

The transformation S maps $\mathcal{W}_s$ onto E_s, $\mathcal{W}_u$ onto E_u, and then $S\,T\,S^{-1}$ is equal to the <u>linear</u> map $\mathcal{D}\,\mathcal{N}_\epsilon(\varphi_\epsilon)$ plus a remainder which leaves E_s, E_u invariant and is small. All these statements hold in a neighborhood of radius $\mathcal{O}(\epsilon^{90})$ of the origin of L_∞.

We would like to eliminate now the non-linear (small) remainder $\mathcal{R}$ by a further coordinate transformation. This is not possible in general, unless the eigenvalues λ_0, λ_1, . . . of $\mathcal{D}T$ satisfy none of the relations

$$\lambda_j = \Pi \lambda_k^{\alpha_k} \quad , \quad \alpha_k \in \{0, 1, 2, \ldots\} \qquad (4.4)$$

except the trivial ones. Of course it is very hard to decide condition (4.4) for the infinitely many eigenvalues of $\mathcal{D}T$. So we shall be less ambitious and try to linearize the map T only in the unstable directions. In fact, for sufficiently small $\varepsilon > 0$ we have

THEOREM 4.3. Sufficiently near to any of the bifurcation points $c = 2^{1/j}$, $j = 2, 3, \ldots$ the eigenvalues $\lambda_{2k}^{(j)}$, $k = 0, 1, \ldots 2(j-1)$ do not satisfy any relation of the form (4.4). (Proof: cf Section 8.)

Therefore, one can linearize the flow on the even subspace of L_∞ in the unstable directions and we have the final form of the description of the flow around φ_ε :

THEOREM 4.4. For sufficiently small $\varepsilon > 0$, there is a neighborhood $\mathcal{V}_\varepsilon$ of the origin of L_∞, a C^1 diffeomorphism $\mathcal{U}_\varepsilon$ and a contraction L_ε , (with $\|L_\varepsilon f\|_\infty \leq \|f\|_\infty^{3/2}$) such that

$$\mathcal{U}_\varepsilon T \mathcal{U}_\varepsilon^{-1} = \left. D \mathcal{N}_\varepsilon(\varphi_\varepsilon) \right|_{E_u^{even}} \oplus \left(\left. D \mathcal{N}_\varepsilon(\varphi_\varepsilon) \right|_{E_s^{even}} + L_\varepsilon \right) .$$

See the note on page 135.

In other words, the map can be made to coincide with the tangent map in the unstable directions on the even subspace. The above construction is explicit and can in principle be calculated in perturbation theory. In the literature on the RG, the tangent vectors to the unstable manifold away from φ_ε are called the relevant scaling fields and thus the above procedures allow to compute the higher order corrections to the scaling fields as a function of $\varphi_\varepsilon \in L_\infty$ (or $-\log \varphi_\varepsilon$ which would be called the "Hamiltonian").

Unfortunately, the Theorem 4.3 is not true on the even and odd subspaces together, because there is always at least one relation of the form (4.4) satisfied. Indeed, one can easily check that if $\mathcal{N}_\varepsilon(\varphi) = \varphi$, then for $D\mathcal{N}_\varepsilon(\varphi)$ we find that

$\varphi(z)$ is an eigenvector with eigenvalue 2 ,

$z\varphi(z)$ is an eigenvector with eigenvalue $2/c^{\frac{1}{2}}$,

$\partial_z\varphi(z)$ is an eigenvector with eigenvalue $c^{\frac{1}{2}}$.

Therefore, the first eigenvalue is the product of the second and third eigenvalue.

In the second half of Section 5 we shall need a normalized version of $\mathcal{N}$. For completeness, we state already now the analog of Theorem 4.4, for $\widehat{\mathcal{N}}$ but the remainder of this section can be skipped at first reading. The normalized version of $\mathcal{N}$ is written $\widehat{\mathcal{N}}$,

$$\widehat{\mathcal{N}}_\varepsilon(\varphi)(z) = \frac{\mathcal{N}_\varepsilon(\varphi)(z)}{(c/4\pi)^{\frac{1}{2}} \int dz \, \exp(-z^2/2) \, \mathcal{N}_\varepsilon(\varphi)(z)} \qquad , \quad (4.5)$$

see the next section for the motivation of the choice of $\widehat{\mathcal{N}}_\varepsilon$. Of course, $\widehat{\mathcal{N}}_\varepsilon$ is not very different from $\mathcal{N}_\varepsilon$: if $\mathcal{N}_\varepsilon(\varphi) = \varphi$, then there is a constant $\neq 0$ such that $\widehat{\mathcal{N}}_\varepsilon(\hat{\varphi}) = \hat{\varphi}$ and $\hat{\varphi} = \text{const.} \quad \varphi$. Also $\mathcal{D}\widehat{\mathcal{N}}_\varepsilon(\hat{\varphi})$ has the same spectrum as $\mathcal{D}\mathcal{N}_\varepsilon(\varphi)$, except for the eigenvalue 2 of $\mathcal{D}\mathcal{N}_\varepsilon(\varphi)$ which becomes 0 for $\mathcal{D}\widehat{\mathcal{N}}_\varepsilon(\hat{\varphi})$, due to the normalization. In fact, with $d\mu(z) = (4\pi/c)^{\frac{1}{2}}dz \, \exp(-z^2/2)$, $\varphi = \varphi_\varepsilon$, one has

$$\mathcal{D}\widehat{\mathcal{N}}_\varepsilon(\hat{\varphi})f = \mathcal{D}\mathcal{N}_\varepsilon(\varphi)f - \varphi \cdot \int d\mu(z) \quad \mathcal{D}\mathcal{N}_\varepsilon(\varphi)f(z) \Big/ \int d\mu(z)\varphi(z)$$

and if $\mathcal{D}\mathcal{N}_\varepsilon(\varphi)f = \lambda f$, then for

$$g = f - \varphi \cdot \int d\mu(z)f(z) \Big/ \int d\mu(z)\varphi(z)$$

one has $\mathcal{D}\widehat{\mathcal{N}}_\varepsilon(\hat{\varphi})g = \lambda \, g$.

We now state a Corollary of Theorem 4.3 for the case of the

bifurcation from $c = 2^{\frac{1}{2}}$, which is a trivial consequence of Theorem 4.3 and the fact that 2 is not an eigenvalue of $D\widehat{\mathcal{N}}_\varepsilon(\widehat{\varphi}_\varepsilon)$.

COROLLARY 4.5. The eigenvalues λ_1, λ_2, λ_3 of $D\widehat{\mathcal{N}}_\varepsilon(\widehat{\varphi})$ do not satisfy a relation of the form (4.4) and hence the flow defined by $\widehat{T}f = \widehat{\mathcal{N}}_\varepsilon(\widehat{\varphi}_\varepsilon + f) - \widehat{\varphi}_\varepsilon$ can be linearized on the unstable even and odd subspace and one has

$$\widehat{\mathcal{U}}_\varepsilon \, \widehat{T} \, \widehat{\mathcal{U}}_\varepsilon^{\ -1} = D\,\widehat{\mathcal{N}}_\varepsilon(\widehat{\varphi}_\varepsilon)\Big|_{\widehat{E}_u} \oplus (D\,\widehat{\mathcal{N}}_\varepsilon(\widehat{\varphi}_\varepsilon)\Big|_{\widehat{E}_s} + \widehat{L}_\varepsilon) \ .$$

See the note on page 135.

Remarks on Section 4:

The discussion of hyperbolic fixed points (in Banach spaces) can be found in

[25]   M.W. HIRSCH, C.C. PUGH, M. SHUB : Invariant Manifolds. Lecture Notes in Mathematics, Vol. 583, Berlin, Heidelberg, New York. Springer Verlag (1977).

The construction of the normal form is discussed in

[26]   E. NELSON : Topics in dynamics, I. Flows , Mathematical notes, Princeton. Princeton University Press (1969).

The formulae for the corrections to scaling fields are mentioned in

[27]   F.J. WEGNER : Corrections to scaling laws, Phys. Rev. B5, 4529 (1972),

who also gives the conditions (4.4) of Sternberg [26].

5. Discussion of the Critical Indices

One of the triumphs in the RG approach has been the correct prediction of experimentally measured critical indices. The critical indices are defined as follows : Let $Q(\beta)$ be some physical quantity depending on the inverse temperature $\beta = 1/kT$, where k is the Boltzmann constant. Let β_c be some "special" value of β where the physical system under consideration undergoes (possibly) a phase transition, i.e. that some observables exhibit singularities (or diverge) as $\beta \to \beta_c$. Then the <u>critical index of</u> Q <u>at</u> β_c (from above or below) is the limit (if it exists)

$$\nu_Q = \lim_{\beta \to \beta_c} \log Q(\beta) \, / \, \log |\beta - \beta_c| \; .$$

Note that in particular if $\nu_Q < 0$, then this means that $Q(\beta)$ diverges as $\beta \to \beta_c$ and ν_Q measures in some sense how fast this divergence is. As we shall see below, the numbers ν_Q depend on ε , and are called the "trivial indices" or "mean field indices" for $\varepsilon = 0$. One of the tasks of RG-theory is to compute the ν_Q as a function of ε . In this section we show how the critical indices are obtained from the results of Section 4 .

We define now the class of models for which the critical indices can be computed. These models are characterized by the fact that the one-spin part f in the Hamiltonian $\mathcal{K}_{N,f}$ (Eq. 2.1) is a function which is <u>near</u> to $-\beta^{-1} \log \mathcal{S}_\beta(\varphi_\varepsilon)$, the transform of φ_ε into the temperature dependent formulation (cf. 3.3). For other models, we cannot discuss the critical indices, because their mean-spin distributions might not fall into the neighborhood of φ_ε , which is the only region in which we have "perfect" control . In fact, functions which are not near to φ_ε may belong to another universality class, (i.e. domain of attraction of $\mathcal{N}$) and would lead to models with possibly different critical indices. This does not exclude that the Hierarchical Model can be well defined for other one spin parts f, by different methods. However, for a discussion of the RG behaviour, we need this nearness to the critical point φ_ε .

The procedure to define a model is as follows :

1) Fix $\varepsilon > 0$ sufficiently small so that the discussion of Section 4 applies.

2) Choose a number $\beta_{crit} > 0$ which will be the critical temperature of the model we are going to define. For reasons which will become clear later one has to impose

$$\frac{4\pi}{c} \; \frac{2-c}{\beta_{crit}\,^c} \neq \frac{1}{e} \;\; , \;\; c = 2^{\frac{1}{2}(1-\varepsilon)} \;\; . \qquad (5.1)$$

3) Choose a function $\varphi_0 \in \mathcal{W}_s$ (the stable manifold of T , cf. Fig.5). This function should satisfy furthermore the following conditions.

c1) $\varphi_0 > 0$,

c2) $\varphi_0 \in C^1$,

c3) $z \, \partial_z \, \varphi_0(z) \, / \, (\varphi_0(z))^{\frac{1}{2}} \in L_\infty$,

c4) $\log \varphi_0(z) \; z \, \partial_z \, \varphi_0(z) \, / \, (\varphi_0(z))^{\frac{1}{2}} \in L_\infty$,

c5) $\|\partial_z \, \varphi_0 - \partial_z \, \varphi_\varepsilon \|_\infty$ is small .

LEMMA 5.1.   There are on $\mathcal{W}_s$ functions $\varphi_0 \neq \varphi_\varepsilon$ satisfying c1)-c5) (Proof : Section 16).

The choice of β_{crit} and φ_0 determines a Hamiltonian $\mathcal{H} = \mathcal{H}_{N,f_0}$ through the formula

$$\exp(-\beta_{crit}f_0) = \varphi_0(z \left(\frac{\beta_{crit}\,^c}{2-c} \right)^{\frac{1}{2}}) \; \exp(-\beta_{crit} \frac{c}{2-c} \, z^2)(\frac{4\pi}{c} \cdot (\frac{2-c}{\beta_{crit}} \,))^{-\frac{1}{2}}$$

$$= \, S_{\beta_{crit}}(\varphi_0) \; . \qquad (5.2)$$

For any such Hamiltonian we shall calculate the critical indices. They do not depend on the particular choice of φ_0 and this fact is called universality.

Since all our discussions are in terms of the functions φ , we shall describe now how the temperature dependence of the model is reflected in the space of the functions φ . We recall from Section 3 that at inverse temperature β the RG transformation $\mathcal{N}_p^{(\beta)}$ is related to $\mathcal{N}$ through

$$S_\beta^{-1} \exp \{ -\beta \mathcal{N}_p^{(\beta)} (-\beta^{-1} \log S_\beta(F)) \} = \mathcal{N}_\varepsilon(F), \qquad (5.3)$$

where

$$\exp (-\beta \, \mathcal{N}_p^{(\beta)} (f)(z))$$
$$= (2/c^{\frac{1}{2}}) \int ds' \, ds'' \, e^{+\beta \, z^2/2} \, e^{-\beta(f(s')+ f(s''))} \delta(s'+s''-z2c^{-\frac{1}{2}}). \, (5.4)$$

If we define $\varphi(\beta,z) = S_\beta^{-1}(\exp -\beta f_0)$ in extension of Eq.(5.2) then we call the set $\{ \varphi(\beta,.) = S_\beta^{-1}(S_{\beta_{crit}}(\varphi_0))^{\beta/\beta_{crit}} \}$ β near β_{crit}, the temperature trajectory of φ_0 (or f_0).   It is generally assumed in RG theory that the temperature can be used as one coordinate on the unstable manifold $\mathcal{W}_u$. Under the conditions (5.1) and c1), ..., c5) we can show that this is indeed the case.

LEMMA 5.2 . The curve $\varphi(\beta,.)$ is differentiable in L_∞ and it is transversal to the   stable manifold.

Therefore the (inverse) temperature is one of the "relevant" directions and can be used as a coordinate .

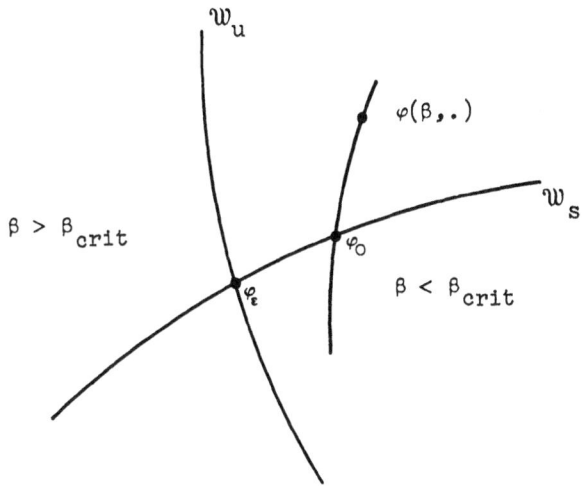

Figure 6. The temperature as a coordinate

We now derive the equations for the moments of the sum of the spins.

Let

$$Z_{N,\beta,f} = \int ds_1 \ldots ds_N \ \exp(-\beta \ \mathcal{H}_{N,f}(s))\qquad(5.5)$$

(the grand canonical partition function for $\mathcal{H}$ in volume N at inverse temperature β) , and let

$$F_{N,\beta,f} = N^{-1} \log Z_{N,\beta,f} \ ,\qquad(5.6)$$

(the free energy). It will be advantageous in the following to distinguish three representations for the function f occurring in $Z_{N,\beta,f}$ and other physical quantities. These are the "physical"(P), "mathematical" (M) and "linearized" (L) representation and they are related as follows :

$$Z_{N,\beta,f}$$

$$= : \overset{P}{Z}_{N,\beta,f} \tag{5.7}$$

$$= \overset{M}{Z}_{N,\beta,S_\beta^{-1}}(\exp(-\beta f)) \tag{5.8}$$

$$= \overset{L}{Z}_{N,\beta,\,\mathfrak{U}_\varepsilon(S_\beta^{-1}(\exp(-\beta f))-\varphi_\varepsilon).} \qquad \text{(cf.Theorem 4.4)} \tag{5.9}$$

Then the action of $\mathcal{N}_\varepsilon$ on even functions is

$$\overset{L}{Z}_{2N,\beta,g}$$

$$= \overset{L}{Z}_{N,\beta,\,(\mathbb{D}\mathcal{N}_\varepsilon(\varphi_\varepsilon)\big|_{E_u}\oplus\mathbb{D}\mathcal{N}_\varepsilon(\varphi_\varepsilon)\big|_{E_s}+L_\varepsilon))\;g} \tag{5.10}$$

$$= \overset{L}{Z}_{N,\beta,\,\mathcal{N}_L^c\;g}. $$

$\mathcal{N}_L^c$ is the normal form of T . The other relations are

$$\overset{M}{Z}_{2N,\beta,g} = \overset{M}{Z}_{N,\beta,\mathcal{N}_\varepsilon^c(g)} \,, \tag{5.11}$$

$$\overset{P}{Z}_{2N,\beta,f} = \overset{P}{Z}_{N,\beta,\mathcal{N}_P^c(\beta)(f)} \,. \tag{5.12}$$

and

$$\mathcal{N}_P^{(\beta)}(f) = -\beta^{-1}\log S_\beta\;\mathcal{N}_\varepsilon\;S_\beta^{-1}(\exp(-\beta f)) \,, \tag{5.13}$$

$$\mathcal{N}_L^c(g-\varphi_\varepsilon) = \mathfrak{U}_\varepsilon(\mathcal{N}_\varepsilon(\varphi_\varepsilon+\mathfrak{U}_\varepsilon^{-1}(g-\varphi_\varepsilon))-\varphi_\varepsilon)$$

$$= \mathfrak{U}_\varepsilon\;T_\varepsilon\;\mathfrak{U}_\varepsilon^{-1}(g-\varphi_\varepsilon) \,, \tag{5.14}$$

$$\mathcal{N}_M^c(g) \;=\; \mathcal{N}_\varepsilon^c(g) \;. \tag{5.15}$$

Before we can go on with the discussion, we want to describe in detail
the limits which occur in this section and in Section 7 . They are
the <u>scaling limit</u> and the <u>thermodynamic limit</u>. Consider as an example the free energy $F_{N,\beta,f}$, but for any other thermodynamic function (as $\chi_{N,\beta,f}$ below) the same definitions apply. The limit

$$\lim_{N \to \infty} F_{N,\beta,f}$$

(if it exists) is called the <u>thermodynamic limit</u> (of the free energy).
Recalling that N always denotes the number of spins, we see that
the thermodynamic limit describes infinite systems. The interest in
describing infinite systems is that the existence of the limit implies
that eventually the number $F_{N,\beta,f}$ does not depend very much on N,
so that the thermodynamic behaviour of large systems is almost independent of their size. In the <u>scaling limit</u> , we assume the existence
of a critical temperature β_c, and we give a specific sequence of
$\beta_N \in \mathbb{R}$ converging to β_c as $N \to \infty$. Then the limit

$$\lim_{N \to \infty} F_{N,\beta_N,f}$$

(if it exists) is called the scaling limit of the free energy. Of
course, this definition depends on the choice of β_N, and we shall see
that a judicious choice allows the determination of the <u>critical in-
dices</u>. These are defined now with more precision than in the beginning
of this section as the double limit

$$\lim_{\beta \to \beta_c} (\log \lim_{N \to \infty} F_{N,\beta,f} \;/\; \log |\beta - \beta_c|) \;.$$

In this section, we shall only discuss the scaling limits. It will follow in Section 7 from the existence of the thermodynamic limit that the double limit above can also be defined and is equal to the scaling limit.

All scaling behaviour can be traced back to Eqs.(2.4). These equations yield from the definition of Z the relation

$$Z^P_{2N,\beta,f} = Z^P_{N,\beta,\mathcal{N}_P(\beta)(f)} \quad , \tag{5.16}$$

and hence by (5.6)

$$F^P_{2N,\beta,f} = \tfrac{1}{2} F^P_{N,\beta,\ \mathcal{N}_P(\beta)(f)} \ . \tag{5.17}$$

We shall study the relation (5.17) by using the normal form of $\mathcal{N}_P(\beta)$. Let e_j, λ_j, $j = 0, 1, 2, 3$ be the j-th eigenvector and eigenvalue of $\mathcal{D}\,\mathcal{N}_\varepsilon(\varphi_\varepsilon)$, respectively. We next <u>choose</u> a sequence of β_N such that the scaling limit, as defined above, of $F^M_{2^N,\beta_N,\varphi(\beta_N,.)}$ exists and <u>such that</u> $\mathcal{N}^N\,\varphi(\beta_N,\cdot)$ <u>stays near</u> φ_ε but is not equal to φ_ε. We shall henceforth only talk about scaling limits if both the above conditions are met. For $\zeta \neq 0$ define $\beta_N = \beta_{crit} + \zeta\lambda_0^{-N}$. Then, by Lemma 5.2,

$$\varphi^L_N = \mathcal{U}_\varepsilon(\varphi(\beta_N,.)) - \varphi_\varepsilon) = \sum_{j=0}^{1} a_{2j}\lambda_0^{-N} e_{2j} + r + r" , \tag{5.18}$$

where $r" = \mathcal{O}(\lambda_0^{-3N/2})\ \epsilon\ E_u$ and $r\ \epsilon\ E_s$, $a_o \neq 0$. Hence, by Theorem 4.4,

$$\mathcal{N}^K_L\ (\varphi(\beta_N,.) - \varphi_\varepsilon) = \mathcal{N}^K_L\ (\delta\varphi_N)$$

$$= \sum_{j=0}^{1} a_{2j}\lambda_{2j}^K\lambda_0^{-N} e_{2j} + r' + 2\ \mathcal{O}(\lambda_0^{-3N/2}) \ .$$

where $\|r'\|_\infty \leqslant \|r\|_\infty^{((3/2)^K)}$.

We see that the linearization pays off, and we get

$$F^M_{2^N, \beta_N, \varphi(\beta_N, \cdot)} = \frac{1}{2^N} F^L_{1, \beta_N, \mathcal{N}^N_L(\delta\varphi_N)}$$

$$= \frac{1}{2^N} F^L_{1, \beta_N, a_o e_o + \mathcal{O}(1)} \qquad \text{as } N \to \infty \; .$$

Now, $F^L_{1, \beta_N, a_o e_o + \mathcal{O}(1)}$ is nothing else than an integral and it depends continuously on $\mathcal{O}(1)$ in L_∞ , and is <u>not zero</u>.

Therefore, we get

$$\lim_{N \to \infty} \frac{\log F^P_{2^N, \beta_N, -\beta^{-1}_{crit} \log s_{\beta_{crit}}(\varphi_o)}}{\log |\beta_N - \beta_{crit}|}$$

$$= \lim_{N \to \infty} \frac{\log \frac{1}{2^N} F^L_{1, \beta_N, a_o e_o}}{\log \lambda^{-N}_o \zeta} = 1 \; . \tag{5.19}$$

Anticipating the existence of the thermodynamic limit, $F^P_{\beta, f}$ $= \lim\limits_{N \to \infty} F^P_{N, \beta, f}$, we get from (5.19) ,

$$\frac{\log F^P_{\beta, f}}{\log |\beta - \beta_{crit}|} \to 1 \quad , \quad \text{as } \beta \to \beta_{crit} \tag{5.20}$$

so that the critical index is 1 in this case.

We shall now do the analogous calculations for the <u>susceptibi-lity</u> and the <u>magnetization</u> with somewhat less details . The suscep-tibility is defined as

$$\chi_{N,\beta,f} = N^{-1} \frac{\int \prod_{j=1}^{N} dS_j \left(\sum_{k=1}^{N} S_k \right)^2 \exp(-\beta \mathcal{H}_{N,f})}{\int \prod_{j=1}^{N} dS_j \exp(-\beta \mathcal{H}_{N,f})} \qquad (5.21)$$

The normalized expectations necessitate a discussion of $\widehat{\mathcal{N}}$, the normalized RG transformation introduced already in Eq. (4.5) .

We describe the quantities for the case of the normalized transformation . They are

$$\widehat{\mathcal{N}}_P^{(\beta)}(g)(z) = \frac{\mathcal{N}_P^{(\beta)}(g)(z)}{\int dz \, \mathcal{N}_P^{(\beta)}(g)(z)} \qquad , \qquad (5.22)$$

$$\widehat{\mathcal{N}}_L(g) = \widehat{u}_\varepsilon \left(\widehat{\mathcal{N}}_\varepsilon(\widehat{\phi}_\varepsilon + \widehat{u}_\varepsilon^{-1} g) - \widehat{\phi}_\varepsilon \right) \qquad (5.23)$$

where $\widehat{u}_\varepsilon$ is the analog of u_ε for $\widehat{\mathcal{N}}_\varepsilon$, and similar symbols are used for S_β , and ϕ

$$\widehat{\mathcal{N}}_M(g) = \widehat{\mathcal{N}}_\varepsilon(g) = \frac{\mathcal{N}_\varepsilon(g)(z)}{(c/4\pi)^{\frac{1}{2}} \int dz \, \exp(-z^2/2) \, \mathcal{N}_\varepsilon(g)(z)}$$

$$(5.24)$$

A substitution of (3.1) into (5.22) shows that (5.22) is consistent with the Eq. (5.24) and the relation between "M" and "P".
The relation analogous to (5.17) is

$$\chi^P_{2N,\beta,f} = \frac{2}{c} \chi^P_{N,\beta,\widehat{N_P}(\beta)(f)} \quad , \tag{5.25}$$

and setting now $\quad \beta_N = \lambda_2^{-N} \zeta + \beta_{crit}$, $\zeta \neq 0$, we get (cf. 5.18)

$$\chi^M_{2N,\beta_N,\varphi(\beta_N,.)} = \left(\frac{2}{c}\right)^N \chi^L_{1,\beta_N,\widehat{N_L}(\delta\widehat{\varphi}_N)}$$

$$= \left(\frac{2}{c}\right)^N \chi^L_{1,\beta_N,a_2\widehat{e}_2 + \mathcal{O}(1)} \quad , \; a_2 \neq 0 . \tag{5.26}$$

Therefore

$$\lim_{N \to \infty} \frac{\log \chi^P_{2^N,\beta_N,-\beta_{crit}^{-1} \log \widehat{s}_{\beta_{crit}}(\widehat{\varphi}_o)}}{\log |\beta_N - \beta_{crit}|}$$

$$= \lim_{N \to \infty} \frac{\log \left(\frac{2}{c}\right)^N \chi^L_{1,\beta_N,a_2\widehat{e}_2}}{\log \lambda_2^{-N} \zeta} = \frac{\log(c/2)}{\log \lambda_2} \quad .$$

Anticipating again the existence of the thermodynamic limit, above the critical temperature

$$\chi^P_{\beta,f} = \lim_{N \to \infty} \chi^P_{N,\beta,f}$$

we get

$$\frac{\log \chi^P_{\beta,f}}{\log |\beta - \beta_{crit}|} \to \frac{\log(c(\varepsilon)/2)}{\log \lambda_2(\varepsilon)} \quad \text{as} \quad \beta \to \beta_{crit} \; .$$

An analogous result will be seen to hold below the critical temperature.

It remains to discuss the magnetization. This is slightly more complicated than the other cases, because two parameters, the inverse temperature β and the magnetic field h can be varied. We define

$$
M^P_{N,\beta,h,f} = N^{-1} \frac{\int \prod_{j=1}^{N} dS_j \; (\sum_{k=1}^{N} S_k) \; \exp(-\beta h \sum_{k=1}^{N} S_k - \beta \mathcal{H}_{N,f})}{\int \prod_{j=1}^{N} dS_j \; \exp(-\beta h \sum_{k=1}^{N} S_k) \; \exp(-\beta \mathcal{H}_{N,f})} .
$$

$$(5.27)$$

By going onto the even and odd subspaces, we can absorb the definition of $-\beta h \sum_{k=1}^{N} S_k$ into a function $\hat{\varphi}(\beta,h,.)$ which will coincide with $\hat{\varphi}(\beta,.)$ if h = 0 and which will be of the form

$$\hat{\varphi}(\beta,h,.) = \hat{\varphi}_\varepsilon + a(\beta-\beta_{crit})h\hat{e}_1 + b(\beta-\beta_{crit})\hat{e}_2 + \text{remainder} , \quad (5.28)$$

if $|h|, |\beta- \beta_{crit}|$ are small, because the function $\hat{e}_1$ is of the form
const. $x \exp(-\varepsilon x^4) + \mathcal{O}(\varepsilon^2)$.
As before, an easy calculation using (5.22) shows that (we omit now the h dependence which is in f)

$$
M_{2N,\beta,f} = c^{-\frac{1}{2}} M_{N,\beta, \hat{\mathcal{N}}_p(\beta)(f)} , \quad (5.29)
$$

where f is not necessarily in the even subspace.
We consider the scaling limits $\beta \to \beta_{crit}$, and $h \to h_{crit} = 0$
and we take the choice

$$\beta_N = \beta_{crit} + \zeta \lambda_2^{-N} = \beta_{crit} + \beta_N'' , \quad \zeta \neq 0 , \qquad (5.30)$$

$$h_N = \lambda_1^{-N} \eta \beta_{crit}^{-1} , \quad \eta \neq 0 . \qquad (5.31)$$

according to the principles we have **stated** in the case of the free energy.
Then

$$\lim_{N \to \infty} \frac{\log M_{2^N, \beta_N, h_N, f}^P}{\log |\beta_N - \beta_{crit}|} = \lim_{N \to \infty} \frac{\log M_{2^N, \beta_N, \hat{\mathcal{N}}^M(\hat{\phi}(\beta_N, h_N, .))}^M}{\log |\beta_N - \beta_{crit}|}$$

$$= \lim_{N \to \infty} \frac{\log \left(c^{-\frac{1}{2}} \right)^N + \log M_{1, \beta_N, a_1 \hat{e}_1 \lambda_1^N \beta_N h_N + a_2 \hat{e}_2 \lambda_2^N \beta_N'}^L}{\log |\beta_N - \beta_{crit}|}$$

$$= \frac{\frac{1}{2} \log c}{\log \lambda_2} , \qquad (5.32)$$

if

$$\log M_{1, \beta_N, a_1 \hat{e}_1 \lambda_1^N \beta_N h_N + a_2 \hat{e}_2 \lambda_2^N \beta_N'}^L = o(N) \qquad (5.33)$$

as $N \to \infty$. We claim this is the case if (5.30), (5.31) hold. Indeed,
the l.h.s. of (5.33), is equal to $\log M_{1, \beta_N, a_1 \hat{e}_1 \eta + a_2 \hat{e}_2 \zeta}^L$. But
$M_{1, \beta_N, a_1 \hat{e}_1 \eta + a_2 \hat{e}_2 \zeta}^L \neq 0$ if $\eta \neq 0$, as the expectation of S in an odd
measure, and it is bounded by continuity. Eq.(5.33) remains true if
$\lambda_1^N \beta_N h_N = \exp(-g(N))$ with $g(N) > 0$, $\frac{g(N)}{N} \to 0$ as $N \to \infty$.

Assume now, as before, that the thermodynamic limit

$$M_{\beta, h, f} = \lim_{N \to \infty} M_{N, \beta, h, f}$$

exists. Assume furthermore that

$$M_{h,f} = \lim_{\beta \to \beta_{crit}} M_{\beta,h,f} \quad \text{exists, and is} \ne 0 \quad (M_{h,f} = \lim_{\zeta \to 0} M_{\beta,h,f}).$$

Then again the equality (5.29) carries over and we get

$$\frac{\log M_{h,f}}{\log h} \rightarrow \frac{\log \lambda_2 \cdot \frac{1}{2} \log c}{\log \lambda_1 \cdot \log \lambda_2} = \frac{\frac{1}{2} \log c(\varepsilon)}{\log(2/c(\varepsilon)^{\frac{1}{2}})} \quad \text{as } h \to 0 ,$$

$$(5.34)$$

since (5.32) is also equal to

$$\lim_{N \to \infty} \frac{\log M^{P}_{2^N, \beta_N, h_N, f}}{\log h_N \cdot \log \lambda_2 / \log \lambda_1} ,$$

and $\lambda_1 = 2c^{-\frac{1}{2}}$. This describes the magnetization at the critical temperature as a function of the magnetic field.

Finally assume that $M_{\beta,f} = \lim_{h \to 0} M_{\beta,h,f}$ exists and is different from zero. This will only be the case below the critical temperature ($\beta > \beta_{crit}$), as we shall see later. Again the equality (5.29) carries over and we get

$$\frac{\log M_{\beta,f}}{\log(\beta-\beta_{crit})} \longrightarrow \frac{\frac{1}{2} \log c(\varepsilon)}{\log \lambda_2(\varepsilon)} \quad \text{as } \beta \to \beta_{crit} \quad (5.35)$$

and this describes the spontaneous magnetization near the critical temperature in the two phase region.

Summarizing, we see that the behaviour of the various quantities near the critical temperature or near the critical field ($h = 0$) is completely controlled by the linearization of the tangent map $D\mathcal{N}_\varepsilon^2(\varphi_\varepsilon)$.

Since we have seen that φ_ε is C^∞ in $\varepsilon \geqslant 0$, and since $\mathcal{DN}(\varphi_\varepsilon)$ depends analytically on φ_ε , the (isolated) eigenvalues of $\mathcal{DN}(\varphi_\varepsilon)$ have asymptotic expansions (actually, $\lambda_0 = 2$, $\lambda_1 = 2/c^{\frac{1}{2}}$, so that this statement is only relevant for λ_2). Therefore the critical behaviour can be arbitrarily well computed, provided ε is sufficiently small.

We summarize the results, adding the standard notation for the critical indices

$$\log F_{\beta,f} \Big/ \log(\beta-\beta_{crit}) \rightarrow \frac{\log 2}{\log \lambda_0(\varepsilon)} = 1 \quad ,$$

$$\log M_{\beta,f} \Big/ \log(\beta-\beta_{crit}) \rightarrow \frac{\frac{1}{2}\log c(\varepsilon)}{\log \lambda_2(\varepsilon)} = \text{"}\beta\text{"} \sim \frac{1}{2} \, ,$$

$$\log M_{h,f} \Big/ \log h \rightarrow \frac{\frac{1}{2}\log c(\varepsilon)}{\log \lambda_1(\varepsilon)} = \text{"}1/\delta\text{"} \sim 1/3 \quad ,$$

$$\log X_{\beta,f} \Big/ \log(\beta-\beta_{crit}) \rightarrow \frac{\log(c(\varepsilon)/2)}{\log \lambda_2(\varepsilon)} = \text{"}-\gamma\text{"} \sim -1 \quad .$$

We have kept the notations $\lambda_0(\varepsilon)$ $(=2)$ and $\lambda_1(\varepsilon)$ $(= 2/c(\varepsilon)^{\frac{1}{2}})$ to show that relations hold between the critical indices, independently of the value of $\lambda_j(\varepsilon)$. In fact one checks,

$$\gamma = \beta(\delta - 1) , \quad \text{for all } c .$$

Another critical index, $\text{"}\eta\text{"}$, describes the behaviour of

$$< s_0 \ s_j >_{2^N,f} = \frac{\displaystyle\int \prod_{i=1}^{2^N} ds_i \ s_0 s_j \exp(-\beta \mathcal{H}_{2^N,f})}{\displaystyle\int \prod_{i=1}^{2^N} ds_i \ \exp(-\beta \mathcal{H}_{2^N,f})} \quad ,$$

as a function of j at the critical temperature as $j \to \infty$. This index does not depend on any eigenvalue. Due to the particular definition of the model, we have for $j < N$

$$< s_0 s_{2j} >_{2^N, f} \quad = \quad < (s_0 + s_1)/2 \, , \, (s_{2j} + s_{2j+1})/2 >_{2^N, f}$$

$$= \quad \frac{1}{c} < s_0 \, s_{2j-1} >_{2^{N-1}, \, \mathcal{N}_P^{(\beta)}(f)} \, , \quad \beta = \beta_{crit} \, .$$

In the scaling limit, we fix $N' = N - j$ and write

$$< s_0 s_{2j} >_{2^{N'+j}, f} \quad = \quad c^{-j} < s_0^2 >_{2^{N'}, \, \mathcal{N}_P^{(\beta)j}(f)} \, .$$

Then we get

$$\lim_{j \to \infty} \frac{\log < s_0 s_{2j} >_{2^{N'+j}, f}}{\log 2^j} \quad = \quad - \log_2 c \, ,$$

since $\mathcal{N}_P^{(\beta)}(f)$ converges to the image of φ_ε.

By definition

$$\lim_{j \to \infty} \frac{\log < s_0 s_j >}{\log j} = " \, 2 - d - \eta \, " \, , \quad \text{so that} \quad "\eta" = 1 + \log_2 c \, .$$

Remarks on Section 5:

Our treatment is a precise version of standard arguments, and, in part, an expansion of the discussion in [18].

6. Global Properties of the Flow

So far, we have regarded the action of the renormalization group as a purely local phenomenon in the space of densities φ. In this section we describe the mathematics of the action of the renormalization group in the large while the next section will be devoted to the physical implications of the global properties of the RG. One should stress at this point that while local non-linear problems have found some systematization in the mathematics literature, this is not the case for the kind of global question we are going to ask. The main reason for this lack of systematization seems to be that the answers are relatively straightforward in principle, but extremely painful in concrete situations. Our methods and proofs reflect this, although we have tried to avoid unnecessary lengths and repetitions.

We shall fix $\epsilon > 0$ and follow the flow defined by $\mathcal{N} = \mathcal{N}_\epsilon^\circ$ in the large by keeping a careful control over error terms, and by choosing suitable representations (which change as we follow the flow). First we have to anticipate somewhat the problems which we are going to solve in the next section by means of the mathematical results of this section. We shall in fact be interested in the thermodynamic limit for the Hierarchical Model for temperatures near to the critical temperature. This means that we are interested in the convergence of the moments of the measure defined by the Gibbs ensemble for this model at some temperature as the volume tends to infinity.

It follows from the discussion of Section 2, Eq.(2.3) and Eq.(2.4) that the probability density for $s_1 + \ldots + s_{2^N} \in [s, s+ds]$ in volume 2^N at inverse temperature β and with (free) single spin distribution φ_0 is given by

$$\frac{ds\ \exp(-\ s^2(c/2)^N/2)\ \mathscr{N}^N(\varphi(\beta,.))((c/2)^{N/2}s)}{\int ds\ \exp(-s^2(c/2)^N/2)\mathscr{N}^N(\varphi(\beta,.))((c/2)^{N/2}s)} \quad . \tag{6.1}$$

(Here, $\varphi(\beta,.)$ is the temperature trajectory of φ, cf. Section 5). When we talk about <u>convergence</u> in the sequel, we mean (and shall prove) <u>convergence for the measure</u> defined by (6.1). In order to study this convergence, it will turn out to be useful to study $\mathscr{N}^N(\varphi(\beta,.))$ when N is small, and to study (6.1) when N is large. In fact, we have found that by studying $\mathscr{N}$ alone, we were unable to produce the bounds necessary to control the convergence of (6.1).

We next describe the results informally. Then we state them more precisely and finally we give "movies" of the different cases, with references to the numbers of the theorems of Part II. Neglecting the "direction" φ_ε in L_∞ , the neighborhood of $\varphi_\varepsilon \in L_\infty$ takes the form shown in Fig. 7.

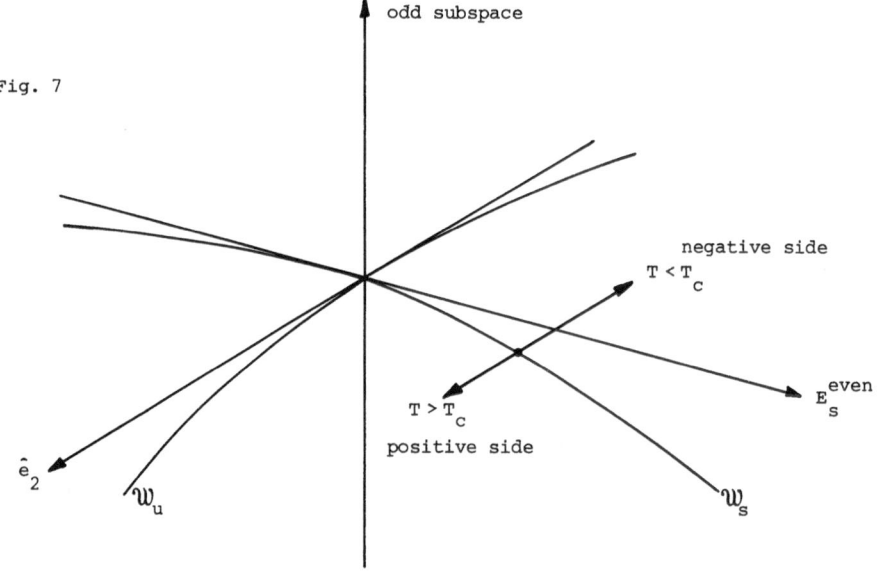

Fig. 7

$\hat{e}_2 \oplus E_s^{even}$ = "zero magnetic field"

Consider a small neighborhood of the origin in this coordinate system. Neglecting for the moment the aspect (6.1), the function f converges under the repeated action of $\mathcal{N}$ to the following limits : (i.e. $\mathcal{N}^n(f)(s)$ converges)

Case 1) to φ_ε if f ϵ $\mathcal{W}_s$: "critical temperature"

Case 2) to a δ function if f $\notin$ $\mathcal{W}_s$, but f ϵ even subspace on the negative side : "zero field, above critical temperature"

Case 3) to two δ functions which separate to infinity if f $\notin$ $\mathcal{W}_s$, but f ϵ even subspace on the positive side of $\mathcal{W}_s$: " zero field, below critical temperature"

Case 4) to one δ-function which moves to $\pm \infty$ if f $\notin$ even subspace : "non-zero field".

In a sense, this means that the "point" $\mathcal{N}^n(\varphi)$ moves from one fixed point to another. This is called CROSSOVER in the literature on critical phenomena. We shall see below that these limits are reached with a "characteristic speed" which is such that in particular the moments defining the susceptibility and the magnetization converge ; in other words, the convergence will be in the sense of (6.1).

The Case 1 above corresponds to the critical temperature and the thermodynamic limit does not exist. However, differently rescaled variables could be made to converge. We now discuss Case 2 in more detail. We first give a collection of more intuitive facts which are of relevance for the mathematical proof of Part II.

i) The "function" δ is a fixed point of $\mathcal{N}$. Indeed,

$$\pi^{-\frac{1}{2}} \int e^{-u^2} \delta(zc^{-\frac{1}{2}} + u) \, \delta(zc^{-\frac{1}{2}} - u) \, du$$
$$= \pi^{-\frac{1}{2}} \, \delta(2c^{-\frac{1}{2}}z) \exp(-z^2/c) = \pi^{-\frac{1}{2}} \, \delta(z) \frac{c^{\frac{1}{2}}}{2} \, ,$$

so that $\delta(x) \cdot \pi^{\frac{1}{2}} 2c^{-\frac{1}{2}}$ is a fixed point of $\mathcal{N}$.

ii) The action of $\mathcal{N}$ on a Gaussian. The operator $\mathcal{N}$ transforms a Gaussian into a Gaussian. Namely

$$\mathcal{N}(\exp(-ax^2))(z)$$

$$= \pi^{-\frac{1}{2}} \int e^{-u^2} e^{-a(zc^{-\frac{1}{2}} - u)^2 - a(zc^{-\frac{1}{2}} + u)^2} du$$

$$= (1 + 2a)^{-\frac{1}{2}} e^{-(2/c)az^2} .$$

The main point is that $a \longrightarrow 2a/c$, so that in some sense the exponential $\exp(-ax^2)$ behaves like the eigenvector H_2 of $\mathcal{D}_2\mathcal{N}(c,1)$ $= \mathcal{A}_{1,\varepsilon}$ cf. page 21 which has eigenvalue $2/c$. This similarity is well known and it is in fact often <u>assumed</u> in the literature that one has $\mathcal{A}_{1,\varepsilon}$ $(\exp - H_2) = \exp(-\mathcal{A}_{1,\varepsilon}(H_2)) = \exp(-2 H_2/c)$, but this fact is only true in first order perturbation theory, and some effort will be needed to bound the difference between the two expressions.

iii) The action of $\mathcal{N}$ on $\exp(-bx^4)$ (a similar discussion applies to $\exp(-ax^2 - bx^4)$). As in Case ii) one calculates easily that

$$\mathcal{N}(e^{-bx^4}) \sim const. \ \exp(-(2/c^2)bx^4)$$

if b is small. In Part II, we shall have to develop the remainder up to order 8 in x.

iv) Near φ_ε, the flow of $\varphi_\varepsilon + f$ is governed by the normal form (Theorem 4.4).

Putting the ideas listed above together, the main stations in the <u>proof of crossover in Case 2</u> are :

(I) Initial point : The function has the form $\varphi_\varepsilon + f, \|f\|_\infty < \varepsilon^{330}$, $\varphi_\varepsilon + f$ on the "negative" side of $\mathcal{W}_s$ (cf. Fig. 7) .

(II) There is an n_1 such that after n_1 iterations, $\mathcal{N}^{n_1} (\varphi_\varepsilon + f) = \varphi^{(n_1)}$ has the form const.$(\varphi_\varepsilon - a\, e_2 + r)$, with $a > \varepsilon^{100}$, $\|r\|_\infty \leqslant \varepsilon^{137}$, $e_2 \sim +(2x^2 \gamma - 1)$ (Corollary 13.2, using ideas of the sort of iv)).

(III) Using detailed estimates on the first few eigenfunctions e_{2j}, we find, by ideas similar to ii), iii), that there is an $n_2 > n_1$ such that

$$\varphi^{(n_2)} = \text{const. } (\varphi_\varepsilon - \varepsilon^{15/16}\, e_2 + \mathcal{O}(\varepsilon^{15/18})\, e_4 + \mathcal{O}(\varepsilon^{23/8})\, e_6$$
$$+ \mathcal{O}(\varepsilon^{31/8})\, e_8 + \mathcal{O}(\varepsilon^{19/8})\, \exp(-\varepsilon \$ x^4/2)),$$

(Theorem 13.7)[*]. This function can be rewritten in a different, more suitable form.

$$\varphi^{(n_2)} = \text{const. } \exp(-\$\varepsilon x^4 - b\varepsilon^{15/16}x^2)\, \{1 + k_6\, \varepsilon^{23/8}x^6 + k_8\varepsilon^{31/8}x^8$$
$$+ \mathcal{O}(\varepsilon^{19/8}\exp(+ \$\varepsilon x^4/2))\}, \text{ (Theorem 14.1).}$$

(IV) Using this new form one finds that there is an $n_3 > n_2$ such that
$$\varphi^{(n_3)} = \text{const. } (\exp(-\varepsilon^{\frac{1}{2}}x^2 - \mathcal{O}(\varepsilon)x^4) + \mathcal{O}(\varepsilon^{11/8}\exp(-\varepsilon^{\frac{1}{2}}x^2)).$$
(Corollary 14.3), where the polynomial terms have disappeared now.

(V) Continuing in this fashion, one can go further and eliminate the quartic term in the exponent, and one gets for some $n_4 > n_3$,
$$\varphi^{(n_4)} = \text{const.}(\exp(-\varepsilon^{1/6}x^2)(1 + \mathcal{O}(\varepsilon^{2/3}))), \text{(Corollary} \quad 14.4).$$

(VI) Now, nothing is in our way to go to $n_5 > n_4$ steps such that
$$\varphi^{(n_5)} = \text{const.}(\exp(-10^4x^2)(1 + \mathcal{O}(\varepsilon^{1/4}))), \text{ (Corollary 14.5).}$$
Note that $n_5 = \mathcal{O}(\log \varepsilon^{-1})$ by Principle ii).

---

[*] We write $\mathcal{O}(a)f(x)$ for short for a term $rf(x)$ with $\|r\|_\infty = \mathcal{O}(a)$.

(VII) In the final step we change to the measure (6.1) instead of studying convergence to a δ-function, and with this in view we rescale the problem by writing a recurrence relation for

$$\mathcal{N}^n(\varphi_\varepsilon + f)((c/2)^{n/2}x) \; .$$

We show then convergence to a Gaussian, (Theorem 14.6).

We give the "movie" of a typical case in Fig. 8a. The figure shows the behaviour of the probability density for the measure (6.1) as a function of the number N, i.e. the number of times the operator $\mathcal{N}$ has been applied. The numerical calculation has been done for an initial probability density (N = 0) which is more concentrated at the origin than the (numerically found) critical spin distribution (N = 0) of Fig. 8b. A detailed interpretation of Figs. 8 and 9 will be given at the end of this section.

The exact theorem in Case 2 is

THEOREM 6.1 . If $\varphi_\varepsilon + f$ is sufficiently near to φ_ε on the high-temperature side of $\mathcal{W}_s$, then

$\mathcal{N}^n(\varphi_\varepsilon + f)(z)$ converges to a δ-function "like a Gaussian" in the sense that

(i)   For some finite constant K,

$$\lim_{n \to \infty} K^{(2^n)}(2/c)^{n/2} \int \exp(-x^2/2)\,\mathcal{N}^n(\varphi_\varepsilon + f)(x)\,dx = C_o \quad \text{exists and}$$

is different from zero.

(ii)   For all m ,

$$\lim_{n \to \infty} \frac{(2/c)^{mn} \int \exp(-x^2/2) \, x^{2m} \, \mathcal{N}^n(\varphi_\varepsilon + f)(x) dx}{\int \exp(-x^2/2) \, \mathcal{N}^n(\varphi_\varepsilon + f)(x) dx} = C_m \quad \underline{\text{exists,}}$$

is different from zero, and $C_m = C_1^{\,m}(2m - 1) !! \,,^*$ i.e. they are the moments of a Gaussian measure.

Discussion of Case 3.

Below the critical temperature, and in zero field, (Case 3) our methods show the separation into two pure phases. We anticipate here again somewhat on the next section to motivate the mathematical question which is going to be asked. The expectation in finite volume 2^N for the variable $s^{2p} 2^{-Np}$ is given by

$$2^{-Np} \frac{\int ds_1 \ldots ds_{2^N} \; (\sum_{j=1}^{2^N} s_j)^{2p} \exp(-\beta_{2^N,f}(s_1, \ldots, s_{2^N}))}{\int ds_1 \ldots ds_{2^N} \; \exp(-\beta_{2^N,f}(s_1, \ldots s_{2^N}))}, \qquad (6.2)$$

which is, according to our discussion in Section 4, equal to

$$\frac{\int ds (2/c)^{Np} s^{2p} \exp(-s^2/2) \, \varphi^{(N)}(s)}{\int ds \, \exp(-s^2/2) \varphi^{(N)}(s)}, \qquad (6.3)$$

where $\varphi^{(N)} = \mathcal{N}^N(\varphi(\beta, .))$. A decomposition into two phases consists now in writing $\varphi^{(N)}(s) = g^{(N)}(s - M_{2^N}) + g^{(N)}(-s - M_{2^N})$, so that by the symmetry $\varphi^{(N)}(s) = \varphi^{(N)}(-s)$, one sees that (6.3) is equal to

---

* $(2m - 1) !! = 1 \cdot 3 \cdot 5 \cdot 7 \ldots (2m - 1).$

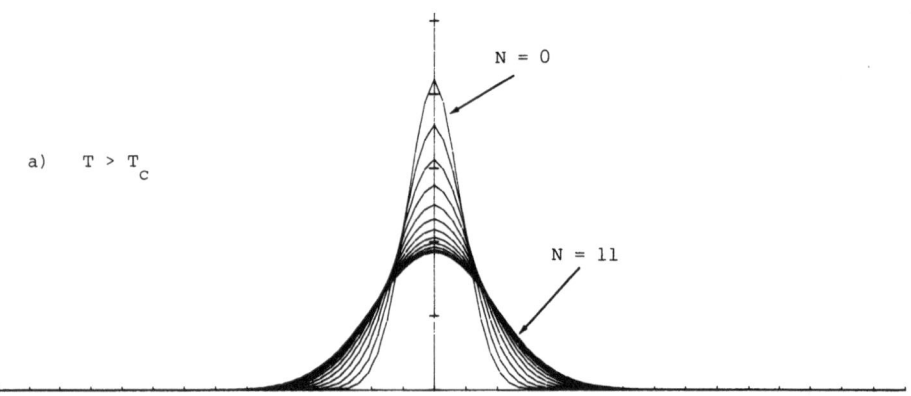

a) $T > T_c$

N = 0

N = 11

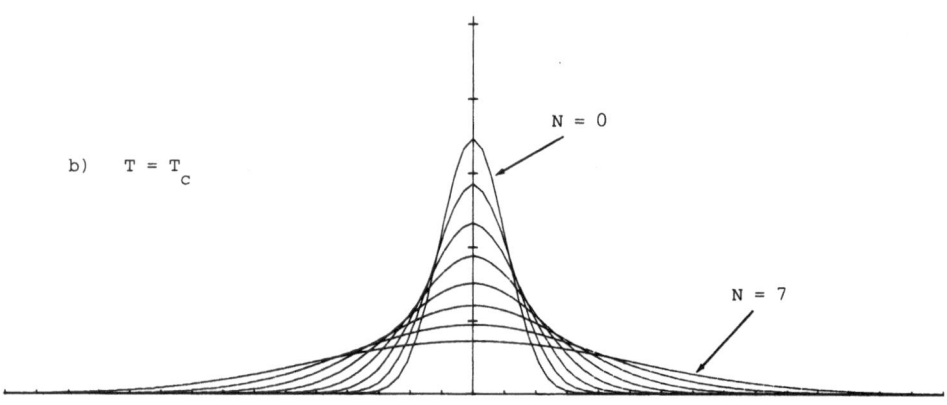

b) $T = T_c$

N = 0

N = 7

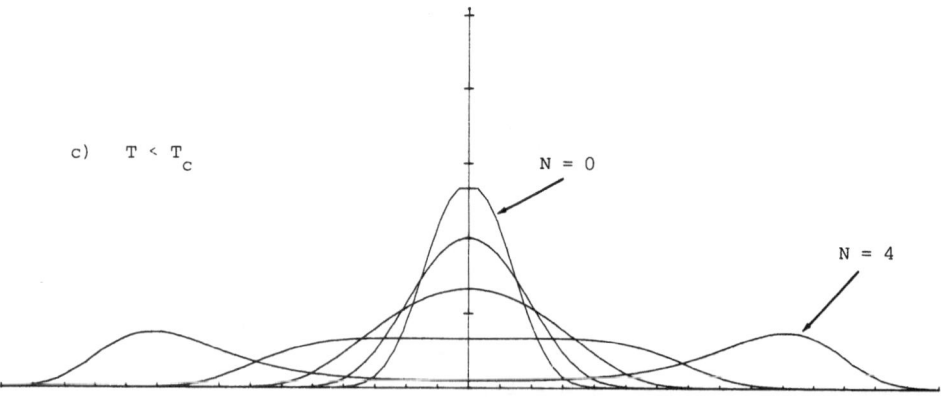

c) $T < T_c$

N = 0

N = 4

Fig. 8.   The spin density for h = 0.

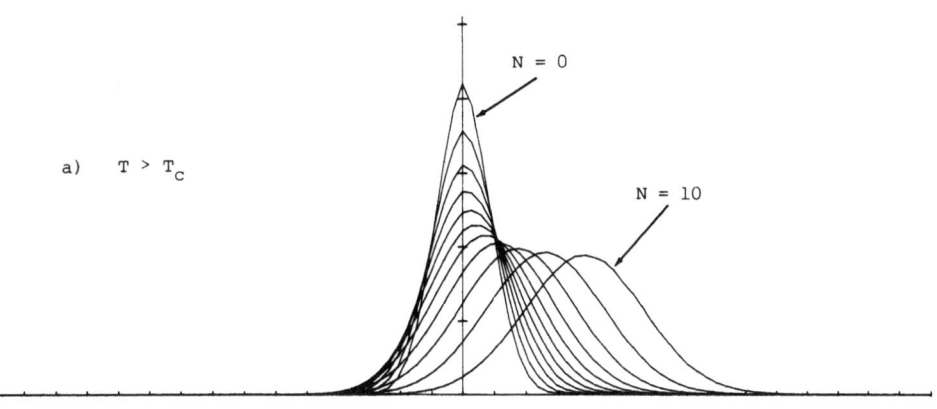

a) $T > T_c$

N = 0

N = 10

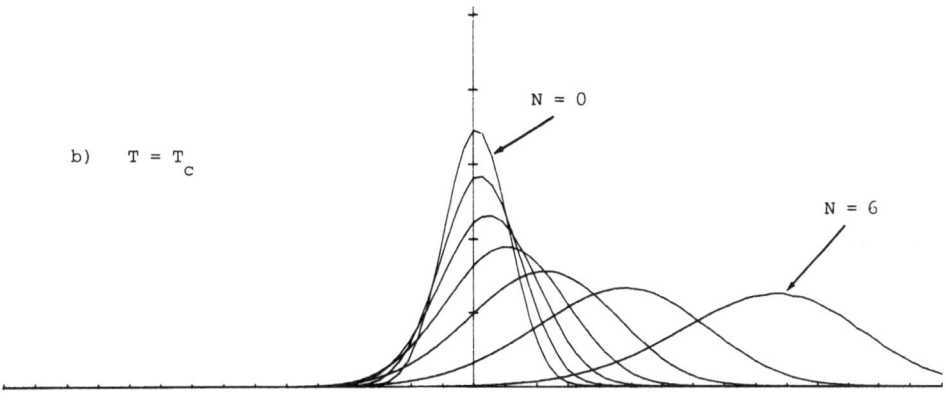

b) $T = T_c$

N = 0

N = 6

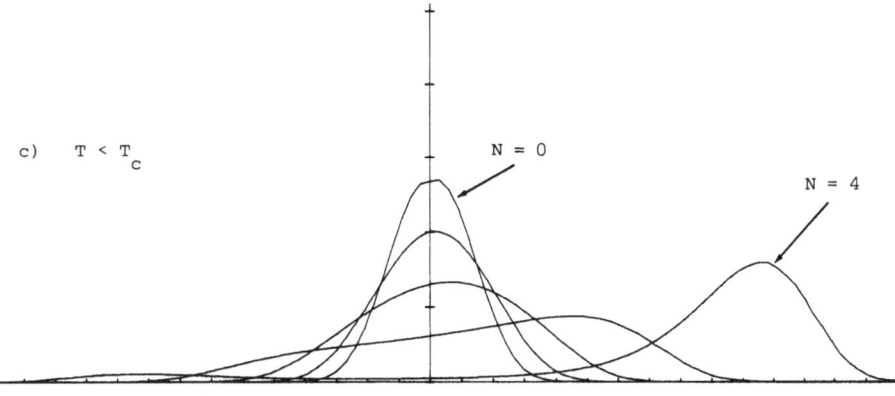

c) $T < T_c$

N = 0

N = 4

Fig. 9. The spin density for h > 0.

$$\tfrac{1}{2} \; \frac{\int ds \; (2/c)^{Np} \; \exp(-s^2/2) \; \exp(-M_{2^N}s) \; g^{(N)}(s)(s + M_{2^N})^{2p}}{\int ds \; \exp(-s^2/2) \; \exp(-M_{2^N}s) \; g^{(N)}(s)}$$

$$+ \; \tfrac{1}{2} \; \frac{\int ds \; (2/c)^{Np} \; \exp(-s^2/2) \; \exp(+M_{2^N}s) \; g^{(N)}(-s) \; (s - M_{2^N})^{2p}}{\int ds \; \exp(-s^2/2) \; \exp(+M_{2^N}s) \; g^{(N)}(-s)} \qquad (6.4)$$

Each of the two terms in (6.4) can be viewed as the expectation of $(s \pm M_{2^N})^{2p}$ in an ensemble described by the weight

$$(2/c)^{Np} \exp(-s^2/2 \mp M_{2^N}s) \; g^{(N)}(s).$$

The aim is to show that <u>there is a choice of $g^{(N)}$</u> such that this en-semble tends to a Gaussian limit as $N \to \infty$.

   This is a very strong result and also painful to prove. In fact it will not only tell us that the system decomposes into two pure (i.e. indecomposable) phases below the critical temperature (a result which is known for many ferromagnetic systems, but not in gene-ral for temperatures arbitrary close to the critical temperature). Our results will suffice to show how the two pure phases form as a function of the volume (and the temperature). This should then also dispel any doubts about the meaning and the utility of studying ther-modynamic limits, since it can be seen here explicitly that already the finite system is (almost) a sum of two systems with spontaneous magnetization. Also, an analysis of the proof, as well as of the computer output Fig. 8c shows that the formation of the two phases (i.e. the decomposition of the measure into two "almost" disjoint "almost" Gaussian measures) occurs relatively suddenly as a function of the volume (i.e. the number of iteration steps). Indeed, we see that a volume of about four times the "critical volume" (i.e. "typical variance") suffices to let the two phases appear as quite distinct maxima. In other words, as soon as the volume exceeds the correlation length (which in turn depends on the temperature) the two phases be - come distinctly visible in the finite system.

We now come to the discussion of some details of the proof. This case is much more complex than Case 2. We use again the ideas mentioned on pp. 61-62 together with some new ideas.

v) The action of $\mathcal{N}$ on a Gaussian which is centered at $b \neq 0$.

$$\mathcal{N}(\exp(-a(x-b)^2))(z) \sim \exp(-(2a/c)(z-bc^{\frac{1}{2}})^2)$$

vi) The function $\exp(-a(x^2-b^2)^2)$ is well approximated by a sum of two Gaussians $\exp(-4ab(x-b)^2) + \exp(-4ab(x+b)^2)$.

We list the main stations in the proof of crossover in Case 3.

(I), (II), (III) are identical to the Case 2. We write again $\varphi^{(n)}$ for $\mathcal{N}^n(\varphi_\varepsilon + f)$.

(IV) After some more steps, one gets

$$\varphi^{(n_3)}(x) = \text{const.}(\exp(\varepsilon^{17/32} x^2 - \mathcal{O}(\varepsilon)x^4) + \mathcal{O}(\varepsilon^{11/8}e^{-\mathcal{O}(\varepsilon)x^4})),$$

(Proposition 15.2).

(V) With some care this representation can be pushed to the point where

$$\varphi^{(n_4)}(x) = \text{const.} \exp(+\varepsilon^{\frac{1}{2}}x^2 - \varepsilon x^4) + \mathcal{O}(\varepsilon^{11/8} e^{-\mathcal{O}(\varepsilon)x^4}),$$

(Proposition 15.3).

(VI) This is rewritten as

$$\exp(-\varepsilon\ x^4 + \varepsilon^{\frac{1}{2}}\ x^2 - 1/4)$$

$$+\ \sum_{\pm}\ \exp(-2\varepsilon^{\frac{1}{2}}\ (\varepsilon^{-1/4}\ 2^{-\frac{1}{2}} \pm x)^2)\ \sigma(\varepsilon^{-1/4}\ 2^{-\frac{1}{2}} \pm x),$$

with $\|\sigma\|_{\mathcal{O}(\varepsilon^{-\frac{1}{2}}),\gamma} = \mathcal{O}(\varepsilon^{11/8})$, and then in Proposition 15.4 we show that this form reproduces, so that we have by Corollary 15.5 the form (for some n_5)

$$\mathcal{N}^{n_5}(\varphi_\varepsilon + f)\ (x)\ \sim\ \exp\ (-\varepsilon\ x^4 +\ \varepsilon^{1/4}\ x^2 - \varepsilon^{-\frac{1}{2}}/4)$$

$$+\ \sum_{\pm}\ \exp(-2\varepsilon^{1/4}(m_5 \pm x)^2)\ \sigma\ (m_5 \pm x) + R(x),$$

with $m_5 = \mathcal{O}(\varepsilon^{-3/8})$, $\|\sigma\|_{\mathcal{O}(\varepsilon^{-1/4}),\gamma} = \mathcal{O}(\varepsilon^{3/4})$, R bounded by a small Gaussian centered at the origin.

This is rewritten in Lemma 15.6 in the form of a double Gaussian

$$\sum_{\pm}\ \exp(-2\varepsilon^{1/4}(m_5 \pm x)^2)\ (1 + \sigma'(m_5 \pm x) + \text{Polynomial})$$

$$+\ R(x).$$

(VII) With two variants of the difficult Proposition 15.4, we get, after n_7 steps in Lemma 15.9, 15.10 to the point where the function $\mathcal{N}^{n_7}(\varphi_\varepsilon + f)(x)$ is of the form

$$\sum_{\pm}\ \exp(-0{,}018(\ m_7 \pm x)^2)\ (1 + \sigma(m_7 \pm x)) + R(x),$$

with $m_7 = \mathcal{O}(\varepsilon^{-\frac{1}{2}})$, $\|\sigma\|_{17,\gamma} = \mathcal{O}(\varepsilon^{1/8})$, R(x) bounded by a small Gaussian, centered at the origin.

(VIII) The function $\mathcal{N}^{n_7}(\varphi_\varepsilon + f)$ has now essentially a fixed form (up to the small error terms) and this reflects the fact that in a suitable scale "all" systems look alike (universality). So from

this point on, we have to deal only with one fixed function
(up to its error terms) and this is the reason why the proofs
become much more "numerical".

Also, in the next 27 steps, the "phase transition" will
take place. In other words, we consider from now on the density
$\exp(-x^2/2) \mathcal{N}^n(\varphi_\varepsilon + f)(x) = \mu_n(x)$. Then $\mu_{n_7}(x)$ looks like one
Gaussian , since it is

$$\exp(-x^2/2) \sum_{\pm} \exp(-0,018(m_7 \pm x)^2)$$

+ errors, while $\mu_{n_7+27}(x)$ looks like two Gaussians , since it
is

$$\exp(-x^2/2) \sum_{\pm} \exp(-(2/c)^{27} \, 0,018(m_7 c^{27/2} \pm x)^2)$$

+ errors.

In fact the transition is very violent and it takes really place
within 2 steps as is also beautifully visible from Fig. 8c ,
N = 2, N = 3, N = 4. The "volume" at which this transition
takes place is thus very well defined, in our case it is
2^{n_7+11} , i.e. depending on the initial function $\varphi_\varepsilon + f$. The
mean spin density will show this sort of transition when there
are 2^{n_7+11} spins, and this number is called the critical volume.
(See also the discussion on page 171).
Our Propositions 15.10 to 15.12 handle this transition.

(IX)   The very painful Theorem 15.13, and its variant 15.14 deal
       then with the separation of the two Gaussians, with a careful
       control of the error terms, using essentially the Principle v)
       of page 69 . Our error terms go then to zero sufficiently fast,
       and, as in the single phase case, with the speed expected, so
       that the system decomposes into two phases in the sense

described in the final result:

THEOREM 6.2.

Let $\varphi_\varepsilon + f \notin W_S$, $\|f\|_\infty \leq \varepsilon^{330}$, $\varepsilon > 0$ <u>sufficiently small</u>, $\varphi_\varepsilon + f > 0$ <u>and even. Suppose further that</u> f <u>is on the side of positive</u> <u>coefficients for</u> $e_2(\simeq + (2x^2 - 1) \exp(-\varepsilon \$ x^4))$. <u>Then</u> $\mathcal{N}^n(\varphi_\varepsilon + f)$ <u>converges to two</u> δ-<u>functions "like a Gaussian" in the following sense:</u>

<u>There is a finite, non-zero constant</u> μ <u>such that</u>

$$(i) \quad \lim_{n \to \infty} 2^{-n} \log \int \exp(-x^2/2) \, \mathcal{N}^n(\varphi_\varepsilon + f)(x) \, dx$$

<u>exists.</u>

$$(ii) \ \underline{\text{One can decompose}} \ \mathcal{N}^n(\varphi_\varepsilon + f)(x) = g_n(x - \mu \, c^{n/2})$$
$+ g_n(-x - \mu \, c^{n/2})$, $\mu \neq 0$, <u>in such a way that the following limits</u>
<u>exist</u> <u>and are different from zero for</u> $p = 0,1,2\ldots$:

$$\lim_{n \to \infty} \frac{(2/c)^{np} \int ds \, \exp(-s^2/2 - \mu \, c^{n/2} \, s) g_n(s) \, s^{2p}}{\int ds \, \exp(-s^2/2 - \mu \, c^{n/2} \, s) \, g_n(s)} \, .$$

<u>They are the moments of a Gaussian measure.</u>

Finally, we discuss the result for the case of a non-zero field which is Case 4. There is thus an asymmetric part in the initial function f, and this asymmetric part will be of the form $\exp(hx)$, $h \neq 0$ which corresponds in physical terms to the case of an external magnetic field. Other forms could be discussed by similar methods but would present less physical interest.

An easy but important observation is that if $f(x) = e^{hx} g(x)$,
then

$$\mathcal{N}_{\varepsilon}(f)(z) = e^{2c^{-\frac{1}{2}} hz} \; \mathcal{N}_{\varepsilon}(g)(z).$$

Therefore, we may discuss the action of $\mathcal{N}$ on the even subspace, and this was done in Case 1 - 3 above. Only for the final convergence a new discussion is necessary. Independently of the fact whether g is on the stable manifold or not, we get the

CONJECTURE 6.3 . For $\|f\|_{\infty} \leqslant \varepsilon^{330}$, $\varepsilon > 0$ <u>sufficiently small and</u> $h \neq 0$ <u>the function</u> $\varphi^{(n)}(z) = \mathcal{N}^n(e^{hx}(\varphi_{\varepsilon} + f)(x))(z)$ <u>converges to a</u> <u>Gaussian in the sense that</u>

(i) $\quad \lim\limits_{n \to \infty} \dfrac{1}{2^n} \log \displaystyle\int e^{-x^2/2} \varphi^{(n)}(x) \; dx \quad$ <u>exists</u>.

(ii) $\quad$ <u>For some constant</u> $\mu \neq 0$, <u>the limits</u>

$$\lim\limits_{n \to \infty} \frac{(2/c)^{np} \displaystyle\int ds \, \exp(-s^2/2 - \mu \, c^{n/2} \, s) \, \varphi^{(n)}(s) \, s^{2p}}{\displaystyle\int ds \, \exp(-s^2/2 - \mu \, c^{n/2} \, s) \, \varphi^{(n)}(s)} \; ,$$

<u>exists for all p and are the moments of a Gaussian measure.</u>

<u>Remark</u> : In view of the standards of rigour we have imposed on these Lecture Notes, we have stated the result as a conjecture, since we have not worked out any proofs (in contrast to all statements we make otherwise in this text). But in view of the technology aquired in the proof of Case 2 and 3 there is little doubt that along exactly the same lines one could prove the conjecture. The proof would be lengthy

and would probably not involve any new ideas. In the case of a model
with Ising-type interaction, one would have the information of Conjecture 6.3 directly from the Lee-Yang theorem [29].

We now comment on the computer output Fig.8 and 9. These drawings
represent the probability densities for the mean spin for the temperature T and the external field h indicated. The numbers N refer to
the number of times the nonlinear map $\mathcal{N}$ has acted on the initial
spin distribution (N = 0), or in other words, the number of spins is
2^N. First of all we see that except for the case T ~ T_c and h = 0,
the shape of the probability density tends to one (or two) Gaussians
with small variance, while in the critical case (T ~ T_c, h = 0) this
variance is large. This corresponds to the divergence of the susceptibility at the critical temperature (in zero field). Still in zero
field and for T < T_c we see that already in volume 2^4 = 16 the two
phases start to form, while in the case with nonzero field one of
the two phases is suppressed.

Remarks on Section 6 :

The results of this section have been stated in several papers
of Bleher and Sinai. However, it seems to us that they did not analyse
completely the separation into two pure phases, but rather the evol-
ution under iterates of $\mathcal{N}$ of a single phase, as in our Conjecture 6.3.

A very nice study of the approach to the thermodynamic limit in
the Ising model for temperatures above the critical temperature, i.e.
in the single phase region has been given in

G.GALLAVOTTI, H.J.F. KNOPS : Block spin interactions in the Ising
        model. Commun. Math. Phys. $\underline{36}$, 171 (1974),

and in

G. GALLAVOTTI, A. MARTIN-LÖF : Block spin distributions for short
        range attractive Ising models. Il Nuovo Cimento $\underline{25B}$, 425(1975).

In these papers, asymptotic expansions in the inverse volume (virial
expansions) are obtained. A very complete control of convergence is also
found for another family of interactions, with stronger forces than those
found in the Hierarchical Model, in

R.S. ELLIS, C.M. NEWMAN : Limit theorems for sums of dependent random
        variables occurring in statistical mechanics (Preprint 1977).

7. Discussion of the Thermodynamic Limit

In this section, we are going to prove that the thermodynamic limits exist, given the estimates of the previous section. Our method is somewhat special, and does not follow the standard methods of Griffith's inequalities or the Lee-Yang theorem which may also work in these circumstances. We rather stress the fact that a complete control of the flow defined by the RG is sufficient to show the existence of the thermodynamic limit. This is not hard to see.

We begin with the Case 2 of Section 6, which is the case of "zero external field, above the critical temperature". Let us go back to notation of Section 5. We choose a function φ_0 on the stable manifold $\mathcal{W}_S$, satisfying the conditions c1),...,c5) of Page 45 and in addition $\|\varphi_0 - \varphi_\varepsilon\|_\infty < \varepsilon^{330}$. Then for $\beta - \beta_{crit} < 0$ sufficiently small and negative, one has $\|\varphi(\beta,.) - \varphi_\varepsilon\| \leqslant \varepsilon^{330}$ and $f = \varphi(\beta,.) - \varphi_\varepsilon$ satisfies the hypotheses of Theorem 6.1 , (the crossover theorem in the single phase case).

Still with the notation of Section 5, we have for the free energy $F_{N,\beta,f}$ $= N^{-1} \log Z_{N,\beta,f}$ the formulas $Z_{2N,\beta,g}^M = Z_{N,\beta,\mathcal{N}(g)}^M$ or

$$F_{2^n,\beta,g}^M = 2^{-n} F_{1,\beta,\mathcal{N}^n(g)}^M . \tag{7.1}$$

The first statement of Theorem 6.1 is exactly

$$\lim_{n \to \infty} 2^{-n} F_{1,\beta,\mathcal{N}^n(\varphi(\beta,.))}^M \quad \text{exists}, \tag{7.2}$$

so that we have shown the existence of the free energy $F_{\beta,\varphi(\beta,.)}^M$ (per unit volume) in the thermodynamic limit. We next compute its critcal index. By Eq. (7.1), we have

$$F^M_{\beta, \varphi(\beta, .)} = 2^{-K} F^M_{\beta, \mathcal{N}^K(\varphi(\beta, .))} . \qquad (7.3)$$

For fixed φ_0, $\varphi(\beta, .)$ is of the form $\varphi_\varepsilon + a(\beta - \beta_{crit}) e_0 +$ remainder, where the remainder is a sum of a term in the supplement of e_0 of order $\mathcal{O}(\beta - \beta_{crit})$ and a term of order $\mathcal{O}((\beta - \beta_{crit})^2)$. By the analysis of Sections 4, 5,

$$\mathcal{N}^K(\varphi(\beta, .)) = \varphi_\varepsilon + a \lambda_0^K (\beta - \beta_{crit}) e_0 + \text{remainder.}$$

In order to combine the scaling limit with the thermodynamic limit, we define an integer $K = K(\beta)$ by

$$|\beta - \beta_{crit}| = b(\beta) \lambda_0^{-K(\beta)}, \quad 1 \leqslant b(\beta) < 2 .$$

From Eq. (7.3) we get by substituting $K = K(\beta)$,

$$F^M_{\beta, \varphi(\beta, .)} = 2^{-K} F^M_{\beta, \mathcal{N}^K(\varphi(\beta, .))}$$

$$= 2^{-K} F^M_{\beta, \varphi_\varepsilon - ab(\beta) e_0} + \text{remainder} . \qquad (7.4)$$

By our analysis of the flow around the fixed point, the remainder in (7.4) goes to zero in L_∞ as $\beta \uparrow \beta_{crit}$. Therefore $\varphi_\varepsilon - abe_0 +$ remainder varies in a bounded set of L_∞ which in bounded away from $\mathcal{W}_s$. It follows thus from the proof of Theorem 6.1 that $\log F^M_{\beta, \mathcal{N}^K(\varphi(\beta, .))} = X(\beta)$ is bounded for β near β_{crit} with $K = K(\beta)$. Therefore

$$\lim_{\beta \uparrow \beta_{crit}} \frac{\log F^M_{\beta, \varphi(\beta, .)}}{\log |\beta - \beta_{crit}|}$$

$$= \lim_{\beta \uparrow \beta_{crit}} \frac{-K(\beta) \log 2 + \log F^M_{\beta, \mathcal{N}^K(\varphi(\beta, .))}}{-K(\beta) \log \lambda_0 + \log b(\beta)}$$

$$= \lim_{\beta \uparrow \beta_{crit}} \frac{\log 2 - X(\beta)/K(\beta)}{\log \lambda_0 - b(\beta)/K(\beta)} = 1 \quad,$$

so that the critical index is equal to 1 .

We can discuss now the other observables of Section 5 in exactly the same manner. The susceptibility satisfies the identity

$$\chi^M_{2^K, \beta, \varphi(\beta,.)} = (2/c)^K \chi^M_{1, \beta, \mathcal{N}^K(\varphi(\beta,.))} \tag{7.5}$$

$$= \frac{(2/c)^K \int ds\ s^2\ \exp(-\beta \mathcal{H}_{1, \mathcal{N}_\beta^{(\beta)K}(f)})}{\int ds\ \exp(-\beta \mathcal{H}_{1, \mathcal{N}_\beta^{(\beta)K}(f)})} \quad, \tag{7.6}$$

cf. Eqs.(5.13) - (5.15). Here, f is the function associated to $\varphi_0 \in \mathcal{W}_s$ and β_{crit}, cf. Eq. (5.2). By the definition of the Hamiltonian, $\mathcal{H}_{1,f}$ has no interaction term and we find that (7.6) is equal to

$$(2/c)^K \frac{\int ds\ s^2\ (c/4\pi)^{\frac{1}{2}}\ e^{-s^2/2}\ \mathcal{N}^K(\varphi(\beta,.))(s)}{\int ds\ (c/4\pi)^{\frac{1}{2}}\ e^{-s^2/2}\ \mathcal{N}^K(\varphi(\beta,.))(s)} \quad, \tag{7.7}$$

using Eq. (5.2) and Eq.(5.22), and this is the motivation for Eq.(6.1).

The existence of the limit

$$\lim_{K \to \infty} \chi^M_{2^K, \beta, \varphi(\beta,.)} = \chi^M_{\beta, \varphi(\beta,.)}$$

follows now from Eqs.(7.5)-(7.7) by the second part of Theorem 6.1. Using again Eq. (7.6) and a choice of K(β) such that $\beta = \beta_{crit} - b \frac{\lambda^{-K(\beta)}}{2}$, for $1 \leqslant b < 2$ (because the eigenvector $\hat{e}_0 = \varphi_\varepsilon$ has eigenvalue 0 in normalized transformations $\hat{\mathcal{N}}$), we get

$$\chi^M_{\beta,\varphi(\beta,\cdot)} = (2/c)^{K(\beta)} \chi^M_{\beta,\mathcal{N}^{K(\beta)}(\varphi(\beta,\cdot))} \qquad (7.8)$$

$$= (2/c)^{K(\beta)} \chi^M_{\beta,\varphi_\varepsilon - ab\hat{e}_2 + \text{remainder}} \quad .$$

Therefore, we find

$$\lim_{\beta \uparrow \beta_{\text{crit}}} \log \chi_{\beta,f} \Big/ \log |\beta - \beta_{\text{crit}}| = \frac{\log(c(\varepsilon)/2)}{\log \lambda_2(\varepsilon)} \quad .$$

Finally, the magnetization $M_{\beta,f}$ equals zero in zero field.

We now discuss the analogous questions in <u>Case 3</u> of Section 6, which is the case of "<u>zero external field, below the critical temperature</u>". Some care is needed here in order not to confuse concepts.

The situation described by the flow in the even subspace below the critical temperature corresponds to a spin model whose state is not a pure phase but the superposition of two pure states with spontaneous magnetization. In order to produce pure states the symmetry $s \rightarrow -s$ of the model has to be broken either by boundary conditions (which would change the renormalization group of the model) or by giving an external field which is then decreased to zero after the thermodynamic limit has been taken. This situation would be described by Case 4 which we do not write out in these Lecture Notes. The Case 3 yields results analogous to Case 4 which are usually not found in the literature because the kind of control we have over the function $\mathcal{N}^K(\varphi_\varepsilon + f)$ (and hence over the partition function in a finite volume) is much

more detailed than in general models. In fact, our procedure corresponds to an explicit decomposition of the mixed phase into two pure
phases.

First of all, it is clear from Theorem 6.2 that the free energy
$F_{\beta,f}$ exists in the thermodynamic limit in exactly the same way as
in Case 2. As we have shortly mentioned before, (Eqs. (6.2)-(6.4))the
ensemble described by the thermodynamic limit is in Case 3 a mixture
of two pure phases. In this mixture, the magnetization is zero, and
the susceptibility is infinite. But each of the two pure phases is
"spontaneously" magnetized and has a finite susceptibility which has
the correct scaling behaviour. Also the magnetization has the correct
scaling behaviour.

According to our discussion leading to formula (6.4), the
thermodynamic ensemble given by the Hierarchical Hamiltonian decomposes
into two measures which are Gaussian and which have non-zero mean
(Theorem 6.2). Therefore the spontaneous magnetization

$$M_{\beta,f} = \lim_{K \to \infty} c^{-K/2} M_{\beta, \mathcal{N}_P^{(\beta)K}(f)}$$

exists,is finite,and non-zero. As before, one finds immediately from
the normal form of $\mathcal{N}^K$ the scaling relation

$$\lim_{\beta \uparrow \beta_{crit}} \log |M_{\beta,f}| \; / \; \log |\beta - \beta_{crit}| = \tfrac{1}{2} \frac{\log c(\varepsilon)}{\log \lambda_2(\varepsilon)} .$$

We next look at the susceptibility in a pure phase which is defined
as

$$\chi^{\pm}_{N,\beta,f} = \frac{N^{-1} \int \prod_{i=1}^{N} ds_i \left(\sum_{i=1}^{N} s_i \pm N \, M_{N,\beta,f} \right)^2 \exp(-\beta \mathcal{H}_{N,f})}{\int \prod_{i=1}^{N} ds_i \; \exp(-\beta \mathcal{H}_{N,f})} .$$

The definition with a "+" sign instead of a "-" sign coincides due
to symmetry.

We get the scaling relation

$$\chi^{\pm}_{2N,\beta,f} = (2/c) \; \chi^{\pm}_{N,\beta,\hat{N}_P}(\beta)(f) \qquad , \qquad (7.9)$$

as in Eq. 5.25, by using also Eq. 6.1 . Again Theorem 6.1 yields the
existence of the thermodynamic limit of the susceptibility. This is
a strong result and in fact the existence of the susceptibility for
all temperatures near to but different from the critical temperature is not known in many models. Again, from the scaling relation
Eq. (7.9), we get the existence of the critical index

$$\lim_{\beta \downarrow \beta_{crit}} \frac{\log \chi_{\beta,f}}{\log(\beta-\beta_{crit})} = \frac{\log \, {}^c/2}{\log \lambda_2} \; .$$

It is then obvious from what has been said above (in Case 2 in
particular) that the thermodynamic limits of the quantities described on page 57 exist in this case and have the correct "critical
indices".

The discussion of Case 4, assuming the Conjecture 6.3 is almost
identical to the preceding cases. One just substitutes the definitions, observes that the thermodynamic limit exists, and performs
then the scaling limit to obtain the critical indices.

8. Perturbation Theory

We have seen in Section 3 how the perturbation theory in ε (when $c = 2^{\frac{1}{2}(1-\varepsilon)}$) is built up and a technically detailed application of these prescriptions is given in Section 9 with a computer program.

We now discuss properties of the formal solution to $\mathcal{N}_\varepsilon(\phi) = \phi$ in the form (cf. 3.19)

$$\phi_\varepsilon(x) = \exp(-\varepsilon\,\theta\,x^4)\; P_K(\varepsilon, x) \;+\; \text{remainder} \qquad (8.1)$$

$$= f_{\varepsilon,K} \qquad + \quad \text{remainder},$$

where $P_K(\varepsilon, x)$ is a polynomial of degree K in ε. The existence of a unique polynomial $P_K(\varepsilon, x)$ giving the correct solution follows by multiplying the formal power series for $\phi_\varepsilon(x)$ by $\exp(\varepsilon\,\theta\,x^4)$.

Proof of Lemma 3.5 : We first show that <u>the coefficient of</u> ε^k <u>in P_K</u> <u>is of degree at most $2k$ in x</u>. In fact define[*]

$$\mathcal{M}_\varepsilon(P)(z) = \exp(\varepsilon\,\theta\,z^4)\; \mathcal{N}_\varepsilon\;(\exp(-\varepsilon\,\theta\,x^4)\,P)\,(z)\,, \qquad (8.2)$$

then

$$\mathcal{M}_\varepsilon(f)(z) = \exp(-(2c^{-2}-1)\,\theta\,\varepsilon\,z^4)\,\pi^{-\frac{1}{2}}\Bigg| \; du \; \exp(-u^2-12\varepsilon\theta z^2 u^2/c-2\varepsilon\theta u^4)$$

$$\cdot \; f(z\,c^{-\frac{1}{2}} + u) \;\; f(z\,c^{-\frac{1}{2}} - u)\,. \qquad (8.3)$$

---

[*] We shall always use x as a dummy variable, i.e.
$(\exp(-\varepsilon\theta x^4)P)(y) = \exp(-\varepsilon\,\theta\,y^4)P(y)$, to avoid writing $\exp(-\varepsilon\,\theta\,(\cdot)^4)$.

We may develop the exponential factors (with the exception of $\exp(-u^2)$) as a power series in ε and the remainder will be bounded in magnitude by the Taylor formula : For $A > 0$, one has

$$\exp(-A) \;=\; \sum_{n=0}^{L} (-A)^n/n! \;+\; (-A)^{L+1}/(L+1)! \; \exp(-\theta A) \,, \quad 0 < \theta < 1$$

$$=\; \sum_{n=0}^{L} (-A)^n/n! \;+\; O(A^{L+1}) \;.$$

We find

$$\mathcal{M}_\varepsilon \, (P_K(\varepsilon \,,\, .))\,(z) \;=\; \{1 + \sum_{n=1}^{[K/2]} (-\varepsilon\,\theta\,(2c^{-2}-1)\,z^4\,)^n \,/\, n! \,\} \tag{8.4}$$

$$\cdot \; \pi^{-\frac{1}{2}} \int du \; e^{-u^2} \{1 + \sum_{n=1}^{K} (-12\varepsilon\,\theta\,x^2 u^2/c\,)^n \,/\, n! \,\}$$

$$\cdot \; \{1 + \sum_{n=1}^{K} (-2\varepsilon\,\theta\,u^4\,)^n \,/\, n! \,\}$$

$$\cdot \; P_K(\varepsilon \,,\, zc^{-\frac{1}{2}}+u) \; P_K(\varepsilon \,,\, zc^{-\frac{1}{2}}-u) \;+\; \mathcal{R} \quad,$$

where

$$|\mathcal{R}| \;\leq\; L \sum_{n=K+1}^{2K} \varepsilon^n (1 + z^{2n}) \;.$$

Note now that $2c^{-2} - 1 = O(\varepsilon)$,   so that <u>all terms on the r.h.s. of</u> (8.4) <u>are of the required mixed degree in</u> ε <u>and</u> x provided $P_K(\varepsilon \,,\, x)$ was of the required degree initially. But by definition (8.1) , $P_K(\varepsilon \,,\, x) = 1$ + higher orders. We now solve equation (8.4) by iteration, as will be done for $\mathcal{N}_\varepsilon$ below. The assertion is proved.

We next prove (3.20) and (3.21). By the definition (8.2), and by (8.4),

$$|\mathcal{M}_\varepsilon\,(P_K(\varepsilon \,,\, .))\,(z) - P_K(\varepsilon \,,\, z)| \;\leq\; L' \sum_{j=K+1}^{2K} \varepsilon^j (1 + z^{2j}) \;. \tag{8.5}$$

If $|z| \leq 10\,K\,\varepsilon^{-\frac{1}{4}} \log(1/\varepsilon)$ then the r.h.s. of (8.5) is $O(\varepsilon^{K/2})$.

On the other hand, if $|z| > 10 \, K \, \varepsilon^{-\frac{1}{4}} \log(1/\varepsilon)$, then

$$e^{-\varepsilon\theta z^4/3} \, |\mathcal{M}_\varepsilon(P_K(\varepsilon, \,.\,))(z)| \leq L'' \, e^{-\varepsilon\theta z^4/3} \, (1 + z^{4K})$$

$$(8.6)$$

$$\leq O(\varepsilon^{K/2}) \,,$$

where the first inequality follows from

$$|\mathcal{M}_\varepsilon(P_K(\varepsilon, \,.\,))(z)| \leq \mathcal{N}_\varepsilon(|P_K(\varepsilon, \,.\,)|)(z) \qquad (8.7)$$

and the fact that $\mathcal{N}_\varepsilon$ at most doubles the degree of a polynomial. This proves (3.20), (3.21) and hence Lemma 3.5. We shall need an additional bound on the derivative of $\quad g_\varepsilon = \mathcal{M}_\varepsilon(P_K(\varepsilon, \,.\,)) - P_K(\varepsilon, \,.\,)$. $\qquad (8.8)$

LEMMA 8.1. $\qquad |\partial_x g_\varepsilon(x) \, e^{-\varepsilon\theta x^4/3}| = O(\varepsilon^{(K+1)/2})$.

Proof : It is clear from (8.8) that g_ε is C^∞ in x. For $|x| \leq 10K \, \varepsilon^{-\frac{1}{4}} \log(1/\varepsilon)$ the result follows from perturbation theory, by sacrificing again half the powers in ε to allow for $|x|$ to go near $O(\varepsilon^{-\frac{1}{4}} \log(1/\varepsilon))$. In the exterior region, the result follows by differentiating the r.h.s. of (8.8) and using the exponential factor to bound the power x^{2K+3}.

We now proceed to the

Proof of Theorem 4.3 : We restate the theorem.

Let $\lambda_o^{(m)}$, $\lambda_1^{(m)}, \ldots$ be the eigenvalues near an m-critical point corresponding to a value of $\quad c = c_\varepsilon^{(m)} = (2/(1+\varepsilon))^{1/m}$. (This is not the same parametrization as in the case $m = 2$, $c = 2^{\frac{1}{2}(1-\varepsilon)}$, but it is more useful in our case.) Then there is for sufficiently small $\varepsilon > 0$, depending on $k \geq 2$, no relation of the form

$$\lambda_i^{(m)}(\varepsilon) = \prod_{j=0}^{\infty} (\lambda_j^{(m)}(\varepsilon))^{k_j} \,, \qquad \sum k_j = k \,, \qquad (8.9)$$

provided

either i is even and $0 \leq i < 2m-2$ and $k_j=0$ for odd j

and (k ≤ 3 or (m=2 and k≥2))

or   m=2, $1 \leq i < 2m-1$, $k_o = 0$.

(Probably the theorem holds for all m ≥ 2 and k ≥ 2 .)

     If the relation (8.9) is to hold, it has to hold in particular up to first order in ε for small ε > 0 since the eigenvalues have asymptotic expansions by Theorem 4.1 , and the discussion on page 57.

     By solving the Eq. (3.14), using the formula (3.17), we find

$$\phi_\varepsilon^{(m)} = 1 - \varepsilon \theta_m \psi_{2m,\varepsilon}^{(m)} + O(\varepsilon^2) , \tag{8.10}$$

with

$$\theta_m = \left[\frac{1}{2} \left(2^{\frac{m-1}{m}} - 1 \right)^m \binom{2m}{m} \left(2m! \right)^{\frac{1}{2}} \Big/ m! \right]^{-1} \tag{8.11}$$

and

$$\psi_{k,\varepsilon}^{(m)} (z) = H_k ((1 - 1/c_\varepsilon^{(m)})^{\frac{1}{2}} z) / (2^k k!)^{\frac{1}{2}} \tag{8.12}$$

Therefore, since $\mathcal{D}_2 \mathcal{N}(c , \phi)$ is linear in ϕ , one finds according to standard perturbation theory

$$\lambda_j^{(m)} (\varepsilon) = 2 / c_\varepsilon^{(m) j/2} - \varepsilon \theta_m 2 \left(\psi_{j,\varepsilon}^{(m)} , \mathcal{N}_{c_\varepsilon}^{(m)} (\psi_{2m,\varepsilon}^{(m)} , \psi_{j,\varepsilon}^{(m)}) \right)$$

$$= 2^{(2m-j)/2} (1 + \varepsilon \left(j/2m - 2 \binom{j}{m} / \binom{2m}{m} \right)) + O(\varepsilon^2) . \tag{8.13}$$

Therefore, Eq. (8.9) holds up to first order in ε only if for some i, and some j_p one has

$$2m - i = \sum_{p=1}^k (2m - j_p) , \tag{8.14}$$

and

$$i/2m - 2\binom{i}{m} / \binom{2m}{m} = \sum_{p=1}^{k} (j_p/2m - 2\binom{j_p}{m} / \binom{2m}{m}) \quad . \tag{8.15}$$

Using (8.14), we can replace (8.15) by the more convenient form

$$\binom{2m}{m} - 2\binom{i}{m} = \sum_{p=1}^{k} (\binom{2m}{m} - 2\binom{j_p}{m}) \quad . \tag{8.16}$$

We state the proof for the case m = 2, k arbitrary, the other cases are similar. Then it is easy to see that if (8.14) and (8.16) hold, then by adding the two equations one must also have

$$\sum_{p=1}^{k} j_p^2 = 10(k-1) + i^2 \quad . \tag{8.17}$$

But from (8.14) we get

$$\sum_{p=1}^{k} j_p^2 > (4(k-1) + i)^2 / k > 16k - 32 + 8i ,$$

so that (8.17) cannot hold if $6k \geq 22 + i^2 - 8i$. This proves the assertion for $k \geq 4$. The cases k = 2,3 are handled by a direct inspection of Eqs. (8.14),(8.17).

Note: If the above analysis is carried to order $O(\varepsilon^q)$, k can be chosen as large as $O(\varepsilon^{-q+1})$.

Remarks on Section 8:

A good discussion of formal power series can be found in

[29] D. RUELLE : Statistical Mechanics, Rigorous Results. New York
                 Benjamin (1969) .

The importance of the Sternberg condition (Theorem 4.3) was stressed by Wegner [19].

9. Explicit Perturbation Calculation for the Eigenvalue λ_2

In this section, we discuss a computer program which calculates the second eigenvalue λ_2 of the tangent map to $\mathcal{N}$ at the fixed point ϕ_ε. The basic point is the formula Eq. (3.17), on the even subspace

$$a_{2k} = \sum_{\substack{|n-n'| \leq k \\ n+n' \geq k}} a_{2n}\, a_{2n'}\, \binom{2k}{k+(n-n')}\, (c^k\, (n+n'-k)!)^{-1}. \qquad (9.1)$$

Setting $b_k = a_{2k}$, $k = 1,2,\ldots$; $b_o = a_o - 1$, we get for b_k the relations

$$b_k = \frac{1}{c^k - 2} \sum_{k}^{*} b_j\, b_{j'}\, \binom{2k}{k+j-j'}\, \frac{1}{(j+j'-k)!}\, , \qquad k \neq 2, \qquad (9.2)$$

where $\displaystyle\sum_{k}^{*}$ extends over $\{j,j' \; ; \; |j-j'| \leq k, \; j+j' > k\}$.

The perturbation calculation will be done for the following parametrization of c, which makes some of the intermediary formulas simpler :

$$c = 2^{\frac{1}{2}}\, (1 - 3/4\, \varepsilon)\, .$$

This implies, as is easily checked, that on the bifurcation branch one has

$$b_2 = -\varepsilon + O(\varepsilon^2),$$

$$b_k = O(\varepsilon^{[k+1]/2}), \qquad k \neq 2 \, , \; [\;\;] = \text{integer part,}$$

$$b_o,\, b_1 = O(\varepsilon^2)\, .$$

Substituting a solution which is correct to order ε^n into the r.h.s. of (9.2) produces the solution correct to order ε^{n+1} on the l.h.s. It remains to treat the case $k=2$. Due to our parametrization, one gets a recursive equation (assuming that b_k is correct to order ε^n for $k \neq 2$, and b_2 is correct to order ε^{n-1} and zero in higher orders). It is :

Coefficient of ε^n in b_2

$$= 1/3 \text{ Coefficient of } \varepsilon^{n+1} \text{ in } \sum_{2}^{*} b_j\, b_{j'}\, \binom{4}{2+j-j'} \frac{1}{(j+j'-2)!}$$

$$- 3/8 \text{ Coefficient of } \varepsilon^{n-1} \text{ in } b_2. \tag{9.3}$$

This last equation is essentially due to the fact that $(c^2-2)^{-1}$
$= -(3\varepsilon)^{-1} + O(1).$

Similarly, let the eigenvector associated to λ_2 have coefficients c_k in the basis $\{\rho_{2k}\}$, (3.15)-(3.16), with $c_1 = 1$, $c_k = O(\varepsilon^{[k/2]+1})$, $k \neq 1$. Also define $\lambda_2 = 2/c + \mu$. Then the equations for c_k and μ are

$$c_k = -\frac{2}{\left(\frac{2}{c^k} - \frac{2}{c}\right)c^k} \sum_{k}^{*} b_j c_j \binom{2k}{k+j-j'} \frac{1}{(j+j'-k)!} - \mu\, c_k \left(\frac{2}{c^k} - \frac{2}{c}\right)^{-1} \tag{9.4}$$

$$\mu = 2 \sum_{1}^{*} b_j c_{j'} \binom{2}{j-j'} \frac{1}{(j+j'-1)!} \tag{9.5}$$

It is easy to program this algorithm, for which we give a program in a formula manipulating language, called Symbal. We give first a listing of the program, followed then by some explanations.

```
'BEGIN'.
'NEW' M2,TT,JP,QQ;
M:=8;   PMOD(4):=1; PMOD(6):=9;
PMOD(1):=0;
PMOD(3):=1;
R:=2**(1/2);

TT:= <> 'FORMAL' K; 'EXIT' 2**(K//2)*R**(K'MOD'2)   <>;
```

```
QQ:=<<0:(M2:=M*2):>>:
T:=1:U:=1:QQ(0):=-1:
QQ(2):=1/3:
SMOD(4):=M:
'FOR' K:=1:M2'DO'
'BEGIN' T:=T*(1-3*E/4); U:=U*2:
   'IF' K 'NE' 2 'THEN'
      'BEGIN' X:=(1-T)*(U+TT.(K+2))/(Z:=U-4);
      QQ(K):= (TT.(K)+2)
            *((Y:=1)+('FOR'N:=1:M'SUM'(Y:=X*Y)))/Z:
      'END'
'END':   ..   (C**K-2)**(-1)
```
} 1

```
SMOD(4):=M:
Q:=<<0:M2+1:>>:
Y:=1:
V:=('FOR'N:=1:M'SUM'(Y:=Y*3*E/4))*(R+1);
Q(0):=   (2+R)/2*((Y:=1)+('FOR'N:=1:M'SUM'(Y:=Y*V))) ;
Q(1):=1:
Z:=1:
'FOR' K:=2:M2+1 'DO'
'BEGIN' Z:=Z*(1-3*E/4); T:=1-Z;W:=1:
   X:=-(TT.(K-1)+1)/(1-2**(K-1))*
   ('FOR' N:=1:M'SUM'(W:=W*T));
   Q(K):=   ( (R/2)*(TT.(K-1)+2**(K-1)) ) *(1-3*E/4)/(1-2**(K-1))
   *((Y:=1)+('FOR'N:=1:M'SUM'(Y:=Y*X)) ):
'END':
..(2/C**K-2/C)**(-1)
```
} 2

```
S:=<<0:M2+1:>>:   S(0):=Q(0):
W:=(R/2)*((Y:=1)+('FOR' J:=1:M'SUM'(Y:=Y*3*E/4)));
Z:=-2:
'FOR' K:=1:M2+1'DO' S(K):=Q(K)*(Z:=Z*W);
..(2/C-2/C**K)**(-1)*2/C**K
```
} 3

```
B:=<<0:M2:>>:
'FOR' J:=0:M2 'DO' B(J):=0:
B(2):=-E:
C:=<<0:M2+1:>>:
'FOR' J:=0:M2+1 'DO' C(J):=0:
C(1):=1:
'FOR' N:=1:M'DO'
'BEGIN' SMOD(4):=N:
'IF' N 'NF' 1 'THEN'                  'BEGIN'
   'FOR' L:= 0: (W:=2*N) 'DO'
   'BEGIN' K:=L; 'IF' L= 2 'THEN' K:=W:
      'IF' L = W 'THEN'
         'BEGIN' K:=2;SMOD(4):=N+1;W:=W+2;H:=B(2) 'END';
         B(K):=DELAYED.(QQ(K))*
         ('FOR'J:=K/2:W-2 'SUM'
            ('FOR'JP:=(Z:=MAX.(J-K,K-J)):MIN.(J,W-2*((J+1)//2))  'SUM'
               B(J)*(B(JP)
            *(('IF'JP=Z'THEN'(Y:=BINOMIAL.(2*K,J-JP+K)/FACT.(J+JP-K))
             'ELSE'(Y:=Y*(K+1+J-JP)/((K+JP-J)*(J+JP-K)))  )
            *('IF' J=JP 'THEN' 1 'ELSE' 2 )))))
   'END':
SMOD(4):=N:
B(2):=H+(COEFF.(B(2),E,N+1)-3/8*COEFF.(H,E,N-1))*E**N;
                                 'END':
..         PHI-1
```
} 4

```
(PMOD(1):=1);
N:=N;
'FOR' J:=0:M2'DO' B(J):=B(J);
PMOD(1):=0;


'BEGIN' SMOD(4):=N;
   'FOR' L:=0: 2*N+1 'DO'
   'BEGIN' K:=L;
      'IF' L=1 'THEN' K:=0;
      'IF' L=0 'THEN' K:=1;
      U:=DELAYED.(S(K));
      C(K):=( 'IF' K=1 'THEN' 0 'ELSE' DELAYED.(Q(K))*F*C(K))   +
      ('FOR' J:= 0: 2*N 'SUM'
      ('FOR' JP:=(Z:= MAX.(K-J,J-K)):
                    MIN.(2*N-2*((J+1)//2)+1,2*N-1,K+J)   'SUM'
         C(JP)*('IF'JP=Z'THEN'
                   (Y:=BINOMIAL.(2*K,J-JP+K)/FACT.(J+JP-K)*B(J)*U)
               'ELSE'(Y:=Y*((K+1+J-JP)/((K+JP-J)*(J+JP-K)))) )));
'IF' K=1 'THEN' 'BEGIN' F:=-C(1); C(1):=1 'END';
   'FND'
'END';
.. EIGENVECTOR C. EIGENVALUE = 2/C + F


(PMOD(1):=1);
'FOR' J:=0:M2+1'DO' C(J):=C(J);
F:=F;
G:=F+R*((Y:=1)+('FOR'J:=1:M'SUM'(Y:=Y*3*E/4)));
PMOD(1):=0;
'END'


'FND';
```

General Comments on the Programming Language

The language is Symbal, which is a very powerful formula manipu-

lation language. Its style is similar to Algol and we note some pecu-

liarities. Most importantly, any variable can take a formal expression

as its value, and the arithmetic operations are correctly performed.

Two points .. precede comments. Assignments are performed from left

to right, with assignments in brackets ( ) preceding the others. The

delimiter sum has the meaning of giving as a result the sum (over the

for-loop) of the ensuing expressions. The brackets < > enclose subrou-

tine bodies and QQ : = <<a : b :>> defines a vector with components

QQ(a), QQ(a+1),... QQ(b).

Comments on the Program

(1), (2), (3) compute the formal expressions (as power series in ε) for

$$QQ : \quad (c^k-2)^{-1} \qquad \text{if } k \neq 2, \quad 1/3 \text{ if } k = 2 ;$$

$$Q : \quad (2/c^k-2/c)^{-1} \qquad \text{if } k \neq 1, \quad 1 \text{ if } k = 1 ;$$

$$S : \quad -(2/c^k-2/c)^{-1} 2/c^k \text{ if } k \neq 1, -2/c \text{ if } k = 1 ;$$

respectively. SMOD(4) = n means that calculations are to be done to order n (in all variables, especially ε).

(4) is the iterative computation of the b_k, and

(5) is the iterative computation of the c_k and of μ (which is called f in the program).

The result is, up to order 34 ; $\lambda_2(\varepsilon) = \sum a_n \varepsilon^n$, where a_n is given below.

a_n	n	a_n	n	a_n	n
$.1414213562 \cdot 10^1$	0	$-.7349868963 \cdot 10^{17}$	12	$-.4807168254 \cdot 10^{41}$	24
$.3535533906 \cdot 10^0$	1	$.5601344520 \cdot 10^{19}$	13	$.5927124339 \cdot 10^{43}$	25
$-.2883883476 \cdot 10^1$	2	$-.4484051366 \cdot 10^{21}$	14	$-.7552419932 \cdot 10^{45}$	26
$.3457541709 \cdot 10^2$	3	$.3760590623 \cdot 10^{23}$	15	$.9936588447 \cdot 10^{47}$	27
$-.9523572108 \cdot 10^3$	4	$-.3297064675 \cdot 10^{25}$	16	$-.1348746316 \cdot 10^{50}$	28
$.3362380739 \cdot 10^5$	5	$.3016680750 \cdot 10^{27}$	17	$.1887162204 \cdot 10^{52}$	29
$-.1444774702 \cdot 10^7$	6	$-.2876217400 \cdot 10^{29}$	18	$-.2719758132 \cdot 10^{54}$	30
$.7120351538 \cdot 10^8$	7	$.2853935999 \cdot 10^{31}$	19	$.4034240285 \cdot 10^{56}$	31
$-.3897679024 \cdot 10^{10}$	8	$-.2943683480 \cdot 10^{33}$	20	$-.6154380845 \cdot 10^{58}$	32
$.2323372700 \cdot 10^{12}$	9	$.3152779883 \cdot 10^{35}$	21	$.9649197038 \cdot 10^{60}$	33
$-.1488803934 \cdot 10^{14}$	10	$-.3502745886 \cdot 10^{37}$	22	$-.1553772667 \cdot 10^{63}$	34
$.1016555177 \cdot 10^{16}$	11	$.4032839643 \cdot 10^{39}$	23		

In view of some results obtained for the anharmonic oscillator and for ϕ^4 field theories, a question of interest seems to be the Borel summability of the Taylor series for $\lambda_2(\varepsilon)$. We have neither been able to prove the necessary bounds on the derivatives of the functions nor to prove the necessary analyticity conditions.

Heuristic arguments have been given by Brézin, Le Guillou, Zinn-Justin for the ε-expansion [32], which lead to the necessary bounds.

This should however be confronted with the work of Khuri [31] where he carefully argues that Borel summability cannot hold for the ε-expansion of the ϕ_4^4 field theory. The function $\lambda_2(\varepsilon)$ has been computed by Bleher.

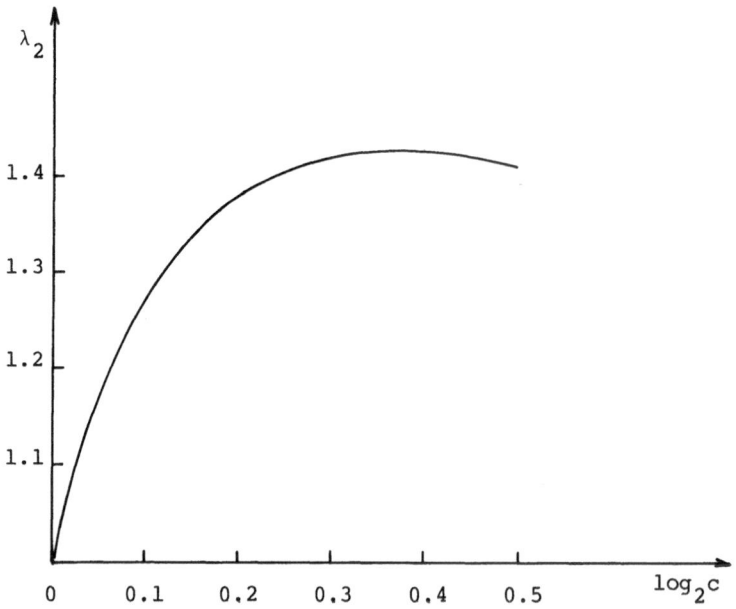

Figure 10. The second eigenvalue as a function of c.

We have done a numerical analysis using the coefficients, together with B. Hirsbrunner. It seems that $a_n \sim$ const. $\Gamma(4,5+n)$ $(3/\log 2)^n$ is a good asymptotic description of the coefficients. One can then compute a function using the resummation methods elaborated by Loeffel [33]. This method allows to compute $\lambda_2(\varepsilon)$ for $c = c(\varepsilon) \in [1, 2^{\frac{1}{2}}]$ and the result agrees with the values given by Bleher [20] which he computed by successive approximations.

The method is based on the following resummation formula, which is absolutely convergent if the original function F has analytic continuations in a two sheeted domain in $|\text{Arg } \varepsilon| \leq 3\pi/2$ with a branch point at $\varepsilon = 0$ and bounds $|\partial_z^n F|(z) \leq n!^2 A^n$:
With $a_n = \partial_z^n F(0)$, one has

$$F(z) = \sum_{n=0}^{\infty} b_n M_n(z) + a_0 ,$$

for $|\text{Arg } z| \leq \pi$, and $|z|$ sufficiently small. Here, b_n, $c_{n,m}$ and M_n are defined as follows.

$$b_n = \sum_{m=0}^{n} c_{n,m} a_{m+1} (\log(2/3))^m / \Gamma(m + 5, 5),$$

$$c_{n,m} = 4^m \binom{m+n-1}{n-m} ,$$

$$M_n(z) = \int_0^{\infty} dt \, e^{-t} t^{4,5} [1 - 2/((1 + 3 tz/\log 2)^{\frac{1}{2}} + 1)]^n.$$

Remarks on Section 9:

A reference for the programming language Symbal can be found in

[30] M.ENGELI . An enhanced Symbal system.SIGSAM Bulletin, ACM $\underline{36}$, 21
     (1975).

We thank the Rechenzentrum der Universität Zürich and the Fides
Trust for computer time and programming assistance.

[31] N.N. KHURI. Borel Summability and the Renormalization Group. Pre-
     print. Rockefeller University.

[32] E.BREZIN , J.-C. Le GUILLOU, J. ZINN-JUSTIN. Perturbation Theory
     at Large Order. I. Phys. Rev $\underline{D15}$, 1544 (1977).

[33] J.J.LOEFFEL . Transformation of an asymptotic series in a conver-
     gent one. Workshop on Pade Approximants. Marseille 1975.

10. Linear Problems and the Existence of ϕ_ε

In this section, we collect most of the estimates on the tangent map $\mathcal{DN}$ and in a sense these estimates are the key to the success in proving existence, differentiability ... of ϕ_ε . We shall adhere to the following notation, writing $\mathcal{A}_{f,\varepsilon} = \mathcal{DN}_\varepsilon(f)$, that is

$$(\mathcal{A}_{f,\varepsilon} g)(z) = 2\pi^{-\frac{1}{2}} \int du \ e^{-u^2} f(zc(\varepsilon)^{-\frac{1}{2}} + u) \ g(zc(\varepsilon)^{-\frac{1}{2}} - u) . \qquad (10.1)$$

We shall also write $\mathcal{A}_f$ if no confusion arises and $\mathcal{A} = \mathcal{A}_{f=1 , \ \varepsilon=0}$, $\mathcal{A}_\varepsilon = \mathcal{A}_{f=1 , \ \varepsilon}$ for the operator discussed in Section 3. The study of $\mathcal{A}_\varepsilon$ starts with Lemma 3.2 which we repeat for convenience .

LEMMA 3.2 . **For** $0 < a < 1$ **the operator** $f \to \pi^{-\frac{1}{2}} \int \exp(-(az-u)^2) \ f(u) du$ **is selfadjoint on** $L_{2,(1-a^2)}$ **and has spectrum** $\{a^n \mid n = 0 , 1 , 2 , \ldots\}$ **with eigenvectors** $H_n((1-a^2)^{\frac{1}{2}} z)$.

Proof : By the definition (3.7) of the Hermite polynomials, we have

$$\sum_{n=0}^{\infty} \frac{t^n}{n!} H_n(x) = e^{+x^2} \sum_{n=0}^{\infty} \frac{(-t)^n}{n!} \partial_x^n e^{-x^2}$$

$$= e^{+x^2} e^{-t\partial_x} e^{-x^2} = e^{x^2} e^{-(x-t)^2} = e^{-t^2 - 2tx}$$

and hence

$$e^{-t^2 - 2t\rho x} = \sum_n \frac{t^n}{n!} H_n(\rho x) .$$

Applying now the convolution operator, we get by Gaussian integration

$$\tfrac{1}{2} \mathcal{A}(e^{-t^2 - 2t\rho \cdot})(z) = \pi^{-\frac{1}{2}} \int e^{-(az-u)^2} e^{-t^2 - 2t\rho u} du$$

$$= e^{-t^2(1-\rho^2)} e^{-2t\rho az} .$$

Setting $\rho^2 = 1 - a^2$ and comparing coefficients, we find

$$\frac{1}{2} \sum_n \frac{t^n}{n!} \, \mathcal{A}(H_n(\rho.))(z) = \frac{1}{2}\mathcal{A}(e^{-t^2 - 2t\rho.})(z)$$

$$= e^{-t^2 a^2 - 2(ta)\rho z} = \sum_{n=0}^{\infty} \frac{(ta)^n}{n!} H_n(\rho z) \, ,$$

for all t. So the spectrum is a^n, with eigenfunctions $H_n((1-a^2)^{\frac{1}{2}}z)$. It remains to be seen that $\mathcal{A}$ is selfadjoint on $L_{2,(1-a^2)}$, but this follows because the $H_n(x)$ are orthogonal on $L_{2,1}$ (integrate by parts) and hence the $H_n((1-a^2)^{\frac{1}{2}} z)$ are orthogonal on $L_{2,(1-a^2)}$ (and form a basis there). This completes the proof.

Since we want to work in the representation $\phi_\epsilon \sim \exp(-\epsilon\theta x^4) \, P_K(\epsilon, x)$ it is more natural to consider instead of $\mathcal{A}_{f,\epsilon}$ the operator

$$g \longrightarrow e^{\epsilon\theta x^4/2} \, \mathcal{A}_{f,\epsilon} \, (e^{-\epsilon\theta x^4/2} \, g) \, .$$

A study of the operator $\mathcal{A}_{f,\epsilon}$ would lead to a result of the form

$$\phi_\epsilon = \exp(-\epsilon\theta x^4) \, P_K(\epsilon, x) + r \, ,$$

with $\|r\|_\infty$ small. But we want to prove a better bound,

$$|r(x)| \leq O(\epsilon^K \exp(-\epsilon\theta x^4/2)) \, ,$$

and hence we consider the above operator, called $\mathcal{B}_{f,\epsilon}$ below. This complicates the notation, but the main philosophy is that $\mathcal{B}_{f,\epsilon}$ is not very different from $\mathcal{A}_{1,\epsilon}$ in $L_{2,\gamma}$.

We define

$$(\mathcal{B}_{f,\epsilon} \, g)(z) = e^{\epsilon\theta z^4/2} \, \mathcal{A}_{f,\epsilon} \, (e^{-\epsilon\theta x^4/2} \, g)(z)$$

$$= 2\pi^{-\frac{1}{2}} \, e^{-\epsilon\theta(2c^{-2} - 1) \, z^4/2}$$

$$\cdot \int du \, \exp(-u^2(1+6\epsilon\theta z^2/c) - \epsilon\theta u^4) \cdot \quad (10.1')$$

$$\cdot \quad \exp(\varepsilon\theta(zc^{-\frac{1}{2}}+u)^4/2) \ f(zc^{-\frac{1}{2}}+u) \quad g(zc^{-\frac{1}{2}}-u) \ ,$$

$$\mathcal{M}_\varepsilon'\ (r)\,(z) \quad = \quad \exp(\varepsilon\theta z^4/2)\ \mathcal{N}_\varepsilon \quad (\exp(-\varepsilon\theta x^4/2)\ r\ )\,(z)$$

$$= \quad \pi^{-\frac{1}{2}} \exp(-(2/c^2-1)\ \varepsilon z^4\theta\,/\,2)\ \cdot$$

$$\cdot \int du \ \exp(-u^2(1+6\varepsilon\theta z^2/c) - \varepsilon\theta u^4)\ r\ (zc^{-\frac{1}{2}}+u)\,r\,(zc^{-\frac{1}{2}}-u).\ (10.2)$$

The importance of the operators $\mathcal{B}_{f,\varepsilon}$ lies in the fact that they provide a __decreasing bound on the solution__ ϕ_ε . In fact, writing

$$\phi_\varepsilon \quad = \quad f \ + \ \exp(-\varepsilon\theta x^4/2)\ r_\varepsilon \ , \qquad\qquad \text{cf (8.1)}$$

the equation $\mathcal{N}_\varepsilon(\phi_\varepsilon) = \phi_\varepsilon$ is transformed to the equation

$$r_\varepsilon \quad = \quad (\mathcal{B}_{f,\varepsilon}-I)^{-1}\{\exp(-\varepsilon\theta x^4/2)\ P_K\ (\varepsilon,x)$$

$$- \exp(-\varepsilon\theta x^4/2)\ \mathcal{M}_\varepsilon\ (P_K(\varepsilon,x)) - \mathcal{M}_\varepsilon'(r_\varepsilon)\} \qquad (10.3)$$

which corresponds to Eq. (3.24), on the Banach space $\exp(-\varepsilon\theta x^4/2)\ L_\infty$, cf. page 82 for the definition of $\mathcal{M}_\varepsilon$. We now __prove__ (in several steps) the counterpart of __Theorem 3.6__. Let K be given. Let $f = f_{K,\varepsilon} = \exp(-\varepsilon\theta x^4)\ P_K$. Then

$$|\exp(+\varepsilon\theta z^4/2)\ f(z)| \ \le \ 2 \ , \qquad\qquad (10.4)$$

for sufficiently small ε , because the coefficient of ε^j in P_K is at most $O(1 + x^{2j})$. The analog of Theorem 3.6 is then the __main estimate__

THEOREM 10.1 . __For__ f __as above__

$$\|(\mathcal{B}_{f,\varepsilon}-1)^{-1} g\|_\infty \ \le \ O(\varepsilon^{-12})\ \|g\|_\infty \ .$$

The main point is that this is a polynomial bound, in particular, $\exp(\varepsilon^{-1})$ would not be a sufficiently good bound.

We first study $\mathcal{B}_{f,\varepsilon}$ itself. The following facts are relatively straightforward.

PROPOSITION 10.2 . The operator $\mathcal{B}_{f,\varepsilon}$ has the following properties (for ε sufficiently small, not depending on K) :

(i) It is compact on L_∞ .

(ii) It is compact on $L_{2,\gamma}$ $(= L_2(\mathbb{R}, \exp(-(1-c^{-1})x^2)(\dot\gamma/\pi)^{\frac{1}{2}} dx))$.

(iii) The first few eigenvalues satisfy

$$\lambda_j = 2^{1-j/4} + O(\varepsilon) \quad , \qquad j = 0, \ldots, 7, \qquad j \neq 4$$

$$\lambda_4 = 1 - \varepsilon\alpha_1 + O(\varepsilon^{3/2}) \quad , \qquad \alpha_1 > 0 \ .$$

(iv) The remaining part of the spectrum is within a distance $O(\varepsilon)$ from the real interval $[0, 1/2]$.

(v) The spectra of $\mathcal{B}_{f,\varepsilon}$ on L_∞ and $L_{2,\gamma}$ coincide.

(vi) $(\mathcal{B}_{f,\varepsilon} - 1)^{-1}$ is a bounded operator on L_∞ .

Before proceeding to the proof, we need the following general estimate. Let $L_{s,\sigma} = L_s(\mathbb{R}, (\sigma/\pi)^{\frac{1}{2}} \exp(-\sigma x^2) dx)$.

LEMMA 10.3 . Let $s, t, r \geq 1$; $\sigma, \tau > 0$, $s \leq t$. If

$$1 - \sigma/s - \tau/t > 0 \quad , \tag{10.5}$$

and

$$\frac{\rho}{r}(1 - \frac{\sigma}{s} - \frac{\tau}{t}) - \frac{1}{c}(\frac{\sigma}{s} + \frac{\tau}{t}) + \frac{4}{c}\frac{\sigma}{s}\frac{\tau}{t} > 0 \quad , \tag{10.6}$$

<u>then one has for</u> $f \in L_{s,\sigma}$ <u>and</u> $g \in L_{t,\tau}$,

$$\int e^{-u^2} | f (zc^{-\frac{1}{2}} + u) g (zc^{-\frac{1}{2}} - u) | du = h(z) \in L_{r,\rho} \qquad (10.7)$$

<u>and</u>

$$\| h \|_{r,\rho} \leq \text{const.} \| f \|_{s,\rho} \| g \|_{t,\tau} . \qquad (10.8)$$

<u>Furthermore, the map</u> $f \to h$, <u>defined by (10.7) for fixed</u> g <u>is compact.</u>

<u>Proof</u> : If $K(z,u)$ is the kernel of an operator K from $L_s(\mathbb{R}, du)$ to $L_r(\mathbb{R}, dz)$, then the operator K is compact if

$$|K|_{r,s} = \left\{ \int du \left[\int dz | K(z,u) |^r \right]^{s'/r} \right\}^{1/s'} < \infty , \quad s' = s/(s-1) ,$$

(an easy generalisation of [34 p. 158]).

Since we work on $L_{s,\sigma}$, we reduce the situation to $L_s(\mathbb{R}, dx)$ by setting $\overset{0}{f}(x) = f(x) \exp(-\sigma x^2/s)$, and similarly for g and h . Then the kernel corresponding to the map $\overset{0}{f} \to \overset{0}{h}$, is

$$K_g(z,u) = e^{-(\rho/r)z^2} \left\{ e^{-(zc^{-\frac{1}{2}} - u)^2} (e^{(t/\tau)(2zc^{-\frac{1}{2}}-u)^2} \overset{0}{g}(2zc^{-\frac{1}{2}}-u)) \right\} e^{(\sigma/s)u^2} .$$

Using the Hölder inequality in z , we bound $|K_g|_{r,s}^{s'}$ by

$$O(1) \int du \left\{ \int dz \left[\exp(-\rho z^2 - r(zc^{-\frac{1}{2}} -u)^2 + r\frac{\tau}{t} (2zc^{-\frac{1}{2}} -u)^2) \right]^{t/t-r} \right\}^{\frac{s'(t-r)}{tr}}$$

$$\cdot \| g \|_{t,\tau}^{s'} \exp (s'(\sigma/s)u^2) .$$

It is now a straightforward matter to evaluate the Gaussian integrals (first the z integral), and this yields the conditions

$$\frac{\rho}{r} > \frac{1}{c} \left[4\frac{\tau}{t} - 1 \right], \quad 1 - \frac{\tau}{t} - \frac{\sigma}{s} > \frac{r}{c} (1 - \frac{2\tau}{t})^2 / (\rho + \frac{r}{c} (1 - \frac{4\tau}{t})^2) ,$$

which after some transformations can be seen to be equivalent to (10.5),

(10.6). This proves the assertion.

Note that (10.7) does <u>not</u> define a continuous map $L_{s,\sigma} \times L_{s,\sigma} \to L_{s,\sigma}$, whatever $s \geq 1$, $\sigma > 0$ may be if $c < 2$. We shall use the following special cases later :

The map (10.7) is compact and continuous on the spaces

$$L_{2,\sigma} \times L_{2,\sigma} \longrightarrow L_{2,3\sigma/c} \ ; \tag{10.9}$$

$$L_{2,\sigma} \times L_{t,\tau} \longrightarrow L_{s,\sigma} \ ,$$

provided $c \geq 6/5$, $\sigma/s \leq 1/8$, $\tau/t \leq \sigma/(20s)$, $s \leq t$.

In particular one can choose

$$L_{2,\sigma} \times L_{4,\tau} \to L_{2,\sigma} \ , \tag{10.10}$$

provided $c \geq 6/5$, $\sigma \leq 1/4$, $\tau \leq \sigma/10$, and

$$L_{4/3,\sigma} \times L_{2,\tau} \longrightarrow L_{4/3,\sigma} \ , \tag{10.11}$$

provided $c \geq 6/5$, $\sigma \leq 1/6$, $\tau \leq 3\sigma/40$.

<u>Remark.</u> Lemma 10.3 holds with $e^{-u^2} du$ replaced by $u e^{-u^2} du$ (with a slight change of the constants, but not of (10.5), (10.6)).

<u>Proof of Proposition 10.2:</u>

(i) Let g be an element in the unit ball of L_∞ .
Then

$$| \mathcal{B}_{f,\varepsilon}(g)(z) | \leq 2 \pi^{-\frac{1}{2}} \exp(-\varepsilon\theta(2c^{-2} -1) z^4/2)$$

$$\cdot \int du \ e^{-u^2} |\exp(-\varepsilon\theta(zc^{-\frac{1}{2}} -u)^4/2) \ P_K(\varepsilon, zc^{-\frac{1}{2}} -u)| \ |g(zc^{-\frac{1}{2}} +u)|$$

$$\leq \|g\|_\infty \ O(\exp(-\varepsilon\theta(2c^{-2} -1) z^4/2)) \ , \tag{10.12}$$

so that the image of the unit ball has uniform decay at infinity. But the image of g is also equicontinuous since $\mathcal{B}_{f,\epsilon}$ is a convolution operator with smooth kernel in $L_1(dz)$. Compactness now follows by the theorem of Arzela-Ascoli.

(ii) This follows by Lemma 10.3 from the first inequality in (10.12).

(iii) In $L_{4,\tau}$, we have

$$\exp(-\epsilon\theta x^4/2)P_K(\epsilon,x) = 1 - \epsilon\theta R_4 + O(\epsilon^{3/2}) , \qquad (10.13)$$

where R_4 is a polynomial of degree 4 which is independent of ϵ . Let D_ϵ be the multiplication operator by the function $\exp(-\epsilon\theta(2c^{-2}-1)x^4/2)$ occurring on the RHS of (10.1). Then with $\|x\|_{2,\gamma} = \|x\|_{L_{2,\gamma} \to L_{2,\gamma}}$, one has

$$\| D_\epsilon - I \|_{2,\gamma} = O(\epsilon^{3/2}) . \qquad (10.14)$$

By Lemma 10.3, (10.13) and (10.14) imply that the operator $g \to C_\epsilon(g)(z)$

$$= 2\pi^{-\frac{1}{2}} \int du \exp(-u^2 - 6\epsilon\theta u^2 z^2 / c)(1 - \epsilon\theta R_4(zc^{-\frac{1}{2}} + u)) g(zc^{-\frac{1}{2}} - u)$$
$$\cdot\exp(-\epsilon\theta u^4)$$

satisfies

$$\| C_\epsilon - \mathcal{B}_{f,\epsilon} \|_{L_{2,\gamma}} = O(\epsilon^{3/2}) . \qquad (10.15)$$

Developing in addition the ϵ-dependent part of the exponential, we find

$$\| C'_\epsilon - \mathcal{B}_{f,\epsilon} \|_{L_{2,\gamma}} = O(\epsilon^{3/2}) , \qquad (10.16)$$

where $C'_\epsilon(g)(z)$

$$= 2\pi^{-\frac{1}{2}} \int du \exp(-u^2)(1 - \epsilon\theta R_4(zc^{-\frac{1}{2}} + u) - 6\epsilon\theta u^2 z^2 / c) g(zc^{-\frac{1}{2}} - u)$$
$$\cdot\exp(-\epsilon\theta u^4) .$$

Proof : $|(C'_\epsilon - C_\epsilon)(g)(z)| \le O(\epsilon^2 \int du \exp(-u^2) g(zc^{-\frac{1}{2}} - u)(1 + z^8 + u^8))$,

because $\exp(-6\epsilon\theta u^2 z^2 / c) - 1 + 6\epsilon\theta u^2 z^2 / c = O(\epsilon^2 \theta^2 u^4 z^4 / c^2)$.

We have thus achieved the comparison of $\mathcal{B}_{f,\varepsilon}$ and C'_ε. But the operator C'_ε is equal to $\mathcal{A}_\varepsilon$ plus a $O(\varepsilon)$ - small perturbation (on $L_{2,\gamma}$). The eigenvalues of $\mathcal{A}_\varepsilon$ are $2/c(\varepsilon)^n$ (Lemma 3.2). To complete the proof of (iii), we have to evaluate the expectation

$$E = - \varepsilon \, 2\pi^{-\frac{1}{2}} \int dz \, \exp(-\gamma z^2) \int du \, \exp(-u^2)$$

$$\cdot \, (\theta R_4 (zc^{-\frac{1}{2}} + u) + 6 \theta u^2 z^2/c + \theta u^4) \, H_4 (\gamma^{\frac{1}{2}}(zc^{-\frac{1}{2}} - u)) \, H_4 (\gamma^{\frac{1}{2}}z)$$

$$\cdot \left[\int dz \, \exp(-\gamma z^2) \, H_4(\gamma^{\frac{1}{2}}z)^2 \right]^{-1} \quad ,$$

which is the correction to the fourth eigenvalue,

$$\lambda_4 = 2 \, / \, c(\varepsilon)^2 + E$$

$$= 1 + \varepsilon \, \log 2 + E + O(\varepsilon^2) \, .$$

One has $R_4(x) = H_4(\gamma^{\frac{1}{2}}x)/32\gamma^2 - 3H_2(\gamma^{\frac{1}{2}}x)/8\gamma^2 - 3/8\gamma^2$. It follows from the Equation 8.13 that the expectation of $H_4/32\gamma^2$ is exactly equal to $-\varepsilon \log 2$, so that it suffices to show that the expectation of the remaining terms is positive. This is proved after some straightforward but tedious calculations using Eqs.(3.8) and (3.14).

(iv) Follows at once from (10.16), and Kato's theorem.

(v) The spectrum of $\mathcal{B}_{f,\varepsilon}$ on L_∞ is discrete by the Riesz–Schauder Theorem. Since $L_\infty \subset L_{2,\gamma}$, it follows that the spectrum of $\mathcal{B}_{f,\varepsilon}$ on L_∞ is contained in the spectrum of $\mathcal{B}_{f,\varepsilon}$ on $L_{2,\gamma}$. On the other hand, because of the decrease of its kernel $\mathcal{B}_{f,\varepsilon}$ maps $L_{2,\gamma}$ into L_∞ so that eigenvalues and eigenfunctions on the two spaces coincide.

(vi) Since 1 is not in the spectrum of $\mathcal{B}_{f,\varepsilon}$ by (iii), (iv), this is immediate.

This completes the proof of Proposition 10.2.

We come now to the <u>main estimate</u> which will guarantee the existence of a solution $\phi_\varepsilon = \mathcal{N}_\varepsilon^\rho(\phi_\varepsilon)$. This is an estimate of $\|(\mathcal{B}_{f,\varepsilon} - I)^{-1}\|_{L_\infty \to L_\infty} =: \|(\mathcal{B}_{f,\varepsilon} - I)^{-1}\|$. Using Proposition 10.2, we denote the first eight eigenfunctions of $\mathcal{B}_{f,\varepsilon}$ on $L_{2,\gamma}$ (even and odd subspace) by $e_0, \ldots, e_7$, and we let $P_0, \ldots, P_7$ denote the corresponding spectral projections (defined through the resolvent formula). Let $P^\perp$ be the projection corresponding to the remainder of the spectrum. Let $E_0, \ldots, E_7, E^\perp$ be the corresponding subspaces of $L_{2,\gamma}$. On L_∞, similar objects can be defined, and they are, by the inclusion $L_\infty \subset L_{2,\gamma}$ and the resolvent formula, the restrictions of the operators or subspaces of $L_{2,\gamma}$ to L_∞. By construction one has

$$(\mathcal{B}_{f,\varepsilon} - 1)^{-1} = \sum_{j=0}^{7} (\lambda_j - 1)^{-1} P_j + (\mathcal{B}_{f,\varepsilon} - 1)^{-1} P^\perp. \tag{10.17}$$

The proof of Theorem 10.1 is based on the following three estimates.

<u>LEMMA 10.4.</u>

$$\|\mathcal{B}_{f,\varepsilon} \, g\|_\infty \leq 4 \, \|g\|_\infty \, .$$

<u>LEMMA 10.5.</u>

$$\|\mathcal{B}_{f,\varepsilon} \, g\|_\infty \leq O(1) \, \|g\|_{p,\gamma} \, , \quad \underline{\text{provided}} \quad p \geq 2\gamma\varepsilon^{-1} + 1 \, .$$

<u>LEMMA 10.6.</u>

$$\underline{\text{For}} \quad q \geq 2 \quad \underline{\text{and}} \quad c = c(\varepsilon) \geq 6/5,$$

$$\|\mathcal{B}_{f,\varepsilon} \, g\|_{(q-1)c+1,\gamma} \leq 4 \, \|g\|_{q,\gamma} \, .$$

Proof of Lemma 10.4 : By the definition (10.1),

$$|(\mathcal{B}_{f,\epsilon}\, g)(z)| \le \exp(-\epsilon\theta(2c^{-2}-1)z^4/2)$$

$$\cdot\ 2\pi^{-\frac{1}{2}}\int du\, e^{-u^2}\left|\exp(-\epsilon\theta(zc^{-\frac{1}{2}}+u)^4/2)P_K(\epsilon,zc^{-\frac{1}{2}}+u)\right|\,|g(zc^{-\frac{1}{2}}-u)|$$

$$\le 2\cdot 2\cdot\|g\|_\infty \qquad , \tag{10.18}$$

by Eq. (10.4).

Proof of Lemma 10.5 : Let $p \ge 2\gamma\epsilon^{-1}+1$ and let $g\in L_{p,\gamma}$.
Using again inequality (10.4) (with a slight variation), we have

$$\left|\exp(-\epsilon\theta x^4/2)P_K(\epsilon,x)\right| \le 2\exp(-\epsilon\theta x^4/4) \quad .$$

Substituting into the definition of $\mathcal{B}_{f,\epsilon}$, we get

$$|(\mathcal{B}_{f,\epsilon}\, g)(z)|$$

$$\le\ 4\pi^{-\frac{1}{2}}\int du\,\exp\left[-u^2(1+6\epsilon\theta z^2/c)-\epsilon\theta(zc^{-\frac{1}{2}}-u)^4/4\right]|g(zc^{-\frac{1}{2}}+u)|$$

$$\le\ 4\pi^{-\frac{1}{2}}\int du\,\exp\left[-(u-zc^{-\frac{1}{2}})^2(1+6\epsilon\theta z^2/c)-\epsilon\theta(2zc^{-\frac{1}{2}}-u)^4/4+\gamma u^2/p\right]$$

$$\cdot\ |\exp(-\gamma u^2/p)\,g\,(u)|$$

$$\le\ O(1)\,\|g\|_{p,\gamma}\left[\int du\,\exp\left[-q\left((u-zc^{-\frac{1}{2}})^2+\theta\epsilon(u-2zc^{-\frac{1}{2}})^4/4-\gamma u^2/p\right)\right]\right]^{\frac{1}{q}},$$

with $p^{-1}+q^{-1}=1$ by the Hölder inequality. It suffices thus to bound
the integral. From $p > 2\gamma\epsilon^{-1}+1$ we have $q\le 2$ and thus
$\gamma q/p < 2\gamma/(2\gamma\epsilon^{-1}+1) < \epsilon$. It suffices therefore to give a uniform
bound in ϵ on the integral :

$$\int \exp(-(y-u)^2)\,\exp(-(\theta/4)\,\epsilon(2y-u)^4)\,\exp(\epsilon u^2)\,du \quad .$$

It is then convenient to divide (for $y \geq 0$) the domain of integration

into two subdomains $|u| \leq y(2-10^{-2})$ and $|u| > y(2-10^{-2})$. In the

first case we have a bound

$$\int \exp(-(y-u)^2) \exp(-(\theta/4)\varepsilon 10^{-8} y^4) \exp(4\varepsilon y^2) \, du = O(1) .$$

In the second case we find $|u - y| > \left[|u| + |y|(1 - 2\cdot10^{-2})\right] / 3$,hence

the bound

$$\int \exp(-(1/9)u^2) \exp(\varepsilon u^2) \, du = O(1) .$$

Proof of Lemma 10.6 : We recall that $\frac{1}{2}\mathcal{A}_{1,\varepsilon}$ is a selfadjoint operator

on $L_{2,\gamma}$, $\gamma = 1 - c(\varepsilon)^{-1}$, and has spectrum $c^{-n/2}$, $n = 0 , 1 , \ldots$

with eigenfunctions $H_n(\gamma^{\frac{1}{2}}x)$, the Hermite polynomials, cf. Theorem 3.1

and Eqs. (3.7), (3.8). Such operators have been extensively studied in

the constructive field theory literature and $\frac{1}{2}\mathcal{A}_{1,\varepsilon}$ is commonly called

the exponential

$$\exp(-\tfrac{1}{2}(\log c) N)$$

of the particle number operator N (in that representation in which the

the ground state wave function is 1), or sometimes one writes

$$\frac{1}{2}\mathcal{A}_{1,\varepsilon} = \Gamma(\exp(-\tfrac{1}{2} \log c)) \qquad (10.19)$$

the second quantization of multiplication by $\exp(-\tfrac{1}{2} \log c)$. Such

operators satisfy the so-called hypercontractive estimates of which the

sharpest form has been given by Nelson :

LEMMA 10.7 . $\frac{1}{2}\mathcal{A}_{1,\varepsilon}$ is a contraction from $L_{q,\gamma}$ to $L_{p,\gamma}$ provided

$q \geq 1$ and

$$\exp(-\tfrac{1}{2} \log c) \leq \left[(q-1) / (p-1)\right]^{\frac{1}{2}} . \qquad (10.20)$$

The proof of this lemma is found in the literature; it is also a consequence of the Sobolev inequalities on $L_p(\mathbb{R})$, and our making use of this fact reflects a similarity between the methods of Bleher-Sinai and Nash-Moser. (See also the bibliographical remarks at the end of this section).

The relation (10.20) is equivalent to

$$p \leq (q - 1) c + 1 .$$

Let now $g \in L_{q,\gamma}$. Then, as in Eq (10.18)

$$| (\mathcal{B}_{f,\varepsilon} \, g) (z) | \leq 2\pi^{-\frac{1}{2}} \exp(-\varepsilon\theta \, (2c^{-2} - 1) \, z^4 / 2)$$

$$\cdot \int du \, \exp(-u^2) \, 2|g(zc^{-\frac{1}{2}} + u)|$$

$$\leq 2(\mathcal{A}_{1,\varepsilon} \, |g|) \, (z) ,$$

so that with $p = (q - 1)c + 1$, we get from (10.20) :

$$\|\mathcal{B}_{f,\varepsilon} \, g\|_{q,\gamma} \leq 2\|\mathcal{A}_{1,\varepsilon} \, |g| \, \|_{q,\gamma}$$

$$\leq 4\| g \|_{(q-1)c + 1,\gamma} .$$

This completes the proofs of Lemma 10.4, 10.5 and 10.6.

We come to the <u>proof of Theorem 10.1</u>. The strategy is as follows. The operators $\mathcal{B}_{f,\varepsilon}$ and in particular their projections P_j are easy to control on $L_{2,\gamma}$. Most importantly, a spectral projection is a bounded operator of norm near 1 on $L_{2,\gamma}$, since $\mathcal{B}_{f,\varepsilon}$ is near $\mathcal{A}_{1,\varepsilon}$ which is selfadjoint on $L_{2,\gamma}$. Such bounds are not true in general on L_∞. In order to bound

$$(\mathcal{B}_{f,\varepsilon} - 1)^{-1} = \sum_{j=0}^{7} (\mathcal{B}_{f,\varepsilon} - 1)^{-1} \, P_j + (\mathcal{B}_{f,\varepsilon} - 1)^{-1}P^{\perp}$$

on L_∞, we go from L_∞ to $L_{2,\gamma}$ and then "come back" from $L_{2,\gamma}$ with the help of Lemma 10.4 - 10.6.

We first bound the P_j, $j = 0, \ldots, 7$. By the above lemmas, we have the following estimate. Let n_o be such that $c^{n_o} \geq 2\gamma\varepsilon^{-1}$, $c^{n_o - 1} < 2\gamma\varepsilon^{-1}$. Then

$$\| P_j\, g \|_\infty = \lambda_j^{-n_o - 1} \| \mathcal{B}_{f,\varepsilon}^{n_o + 1} P_j\, g \|_\infty$$

$$\leq O(1)\, \lambda_j^{-n_o - 1} \| \mathcal{B}_{f,\varepsilon}^{n_o} P_j\, g \|_{c^{n_o} + 1, \gamma} \quad , \text{ by Lemma 10.5}$$

$$\leq O(1)\, \lambda_j^{-n_o - 1}\, 4 \| \mathcal{B}_{f,\varepsilon}^{n_o - 1} P_j\, g \|_{c^{n_o - 1} + 1, \gamma} \quad , \text{ by Lemma 10.6}$$

$$\leq O(1)\, (4 / \lambda_j)^{n_o} \| P_j\, g \|_{2, \gamma}$$

$$\leq O(1)\, (4 / \lambda_j)^{n_o} \| g \|_{2, \gamma} \quad , \text{ since } P_j \text{ is of norm } O(1) \text{ in } L_{2, \gamma}$$

$$\leq O(1)\, (4 / \lambda_j)^{n_o} \| g \|_\infty \quad , \text{ since } L_\infty \subset L_{2, \gamma}\ .$$

Thus, since $\lambda_j > 1/2$ by Proposition 10.2, for $j = 0, \ldots, 7$, we find on L_∞,

$$\| P_j \|_\infty \leq O(1)\, 8^{n_o} \leq O(\varepsilon^{-7})\ . \quad * \qquad (10.21)$$

We proceed to bound $(\mathcal{B}_{f,\varepsilon} - 1)^{-1} P^\perp$ by a Neumann series. Since the sum of projections is 1, we have $\| P^\perp \|_\infty \leq O(\varepsilon^{-7})$. Note now that by Proposition 10.2 and since $\mathcal{B}_{f,\varepsilon}$ is near $\mathcal{A}_\varepsilon$, $\| \mathcal{B}_{f,\varepsilon} P^\perp g \|_{2,\gamma} \leq 3/4 \| g \|_{2,\gamma}$.

---

* No attempt is made at this point to get the best possible estimate.

Therefore, for $n \geq n_o$, we have as before

$$\|\mathcal{B}_{f,\varepsilon}^{n+1} \ P^{\perp}g\|_{\infty} \leq O(1) \ 4^{n_o} \|\mathcal{B}_{f,\varepsilon}^{n-n_o} \ P^{\perp}g\|_{2,\gamma}$$

$$\leq O(1)(3/4)^{n-n_o} \ 4^{n_o} \|P^{\perp}g\|_{2,\gamma}$$

$$\leq O(1)(3/4)^{n+1} \ (16/3)^{n_o} \|g\|_{\infty}$$

$$\leq (3/4)^{n+1} \ O(\varepsilon^{-5}) \|g\|_{\infty} \quad .$$

For $n < n_o$, we have by Lemma 10.4

$$\|\mathcal{B}_{f,\varepsilon}^{n} \ P^{\perp}g\|_{\infty} \leq (3/4)^{n} \ (16/3)^{n} \|P^{\perp}g\|_{\infty} \leq (3/4)^{n} O(\varepsilon^{-12}) \|g\|_{\infty}.$$

Hence

$$\|(\mathcal{B}_{f,\varepsilon} - I)^{-1} g\|_{\infty}$$

$$\leq \sum_{j=0}^{7} (\lambda_j - 1)^{-1} \|P_j \ g\|_{\infty} + \sum_{n=0}^{\infty} \|\mathcal{B}_{f,\varepsilon}^{n} \ P^{\perp}g\|_{\infty}$$

$$\leq \left\{ \sum_{j=0}^{7} (\lambda_j - 1)^{-1} O(\varepsilon^{-7}) + \sum_{n=0}^{\infty} (3/4)^{n} O(\varepsilon^{-12}) \right\} \|g\|_{\infty}$$

$$= O(\varepsilon^{-12}) \|g\|_{\infty} \quad , \quad \text{by Proposition 10.2, (iii).}$$

The proof of Theorem 10.1 is complete.

We proceed to the proof of the existence of ϕ_{ε} .

THEOREM 10.8 . For sufficiently small $\varepsilon > 0$, there is a function $\phi_\varepsilon \neq 1$

solving $\mathcal{N}(\phi_\varepsilon) = \phi_\varepsilon$ and satisfying

$$|\phi_\varepsilon(z)| \leq \exp(-\varepsilon\theta z^4 / 2) \quad \text{for} \quad |z| > (\log 1/\varepsilon)^{\frac{1}{2}} \qquad (10.22)$$

Proof : Consider the Eq. (10.3)

$$r = (\mathcal{B}_{f,\varepsilon} - I)^{-1} \{ \exp(-\varepsilon\theta x^4 / 2) \ P_K (\varepsilon, x)$$

$$- \exp(- \varepsilon\theta x^4 / 2) \ \mathcal{M}_\varepsilon (P_K(\varepsilon, x)) \}$$

$$- (\mathcal{B}_{f,\varepsilon} - I)^{-1} \{\mathcal{M}'_\varepsilon (r_\varepsilon) \}$$

$$= \mathcal{C}_\varepsilon + \mathcal{D}_\varepsilon (r_\varepsilon) . \qquad (10.23)$$

It is easy to see from (10.2) that

$$|\mathcal{M}'_\varepsilon (g) | (z) \leq \mathcal{N}_\varepsilon (|g|) (z) .$$

But $\| \mathcal{N}_\varepsilon (g) \|_\infty \leq \|g\|_\infty^2$,   as follows immediately from the definition.
Thus $\mathcal{D}_\varepsilon$ is a contraction on the ball $\mathcal{K}(\varepsilon^{13}) = \{g \in L_\infty | \|g\|_\infty \leq \varepsilon^{13}\}$,
because $\mathcal{N}_\varepsilon$ is homogeneous of degree 2, and by Theorem lo.1 . On the
other hand, if we choose K = 52 in the definition of P_K ,   then
$\| \mathcal{C}_\varepsilon \|_\infty \leq O(\varepsilon^{14})$,   by a variant of Eq. (lo.3) and by Theorem lo.1 .Thus
the Eq. (10.23) has by the contraction mapping theorem [34, V . 11 . 19]
a unique solution r_ε in $\mathcal{K}(\varepsilon^{13})$ and hence

$$\phi_\varepsilon(z) = \exp(-\varepsilon\theta z^4) \ P_K (\varepsilon, z) + \exp(-\varepsilon\theta z^4 / 2) \ r_\varepsilon (z) \qquad (10.24)$$

is a solution of $\mathcal{N}_\varepsilon (\phi_\varepsilon) = \phi_\varepsilon$,   satisfying the bound (10.22) by a bound
similar to (8.6). Thus Theorem lo.8 is proved.

COROLLARY 10.9 . The function ϕ_ε is once differentiable and

$$|\partial_z \phi_\varepsilon (z)| \leq O(\varepsilon^{\frac{1}{4}}) \exp(-\theta \varepsilon z^4 / 4) \quad .$$

Proof : First of all, $\exp(-c\theta z^4) P_K (\varepsilon, z)$ satisfies the estimate, so it remains to prove the assertion for r_ε . For this we repeat the above arguments on the Banach space S_1 of once differentiable functions with norm

$$\||g\|| = \|g\|_\infty + \|\partial_z g\|_\infty \quad .$$

Integrating by parts in the formula (10.1) defining $\mathcal{B}_{f,\varepsilon}$, we get

$$\partial_z (\mathcal{B}_{f,\varepsilon} g) = c^{-\frac{1}{2}} \mathcal{B}_{f,\varepsilon} \partial_z g + O(\varepsilon^{\frac{1}{4}} \|g\|_\infty)$$

in L_∞ . Therefore, by Theorem 10.1,

$$\partial_z \left((\mathcal{B}_{f,\varepsilon} - 1) g \right) = c^{-\frac{1}{2}} (\mathcal{B}_{f,\varepsilon} - c^{\frac{1}{2}}) \partial_z g + O(\varepsilon^{-12} \| (\mathcal{B}_{f,\varepsilon} - 1) g\|_\infty) ,$$

so that [because $\|\mathcal{B}_{f,\varepsilon} \partial_z g\|_\infty \leq O(\|g\|_\infty) \leq \| (\mathcal{B}_{f,\varepsilon} -1) g\|_\infty O(\varepsilon^{-12})$]

$$\partial_z g = \left((1+\varepsilon^{\frac{1}{2}}) \mathcal{B}_{f,\varepsilon} - c^{\frac{1}{2}} \right)^{-1}$$

$$\cdot \left\{ c^{\frac{1}{2}} \partial_z \left((\mathcal{B}_{f,\varepsilon} -1) g \right) + O(\varepsilon^{-12} \| (\mathcal{B}_{f,\varepsilon} - 1) g\|_\infty) \right\} .$$

But $c^{\frac{1}{2}} / (1+\varepsilon^{\frac{1}{2}})$ is not an eigenvalue of $\mathcal{B}_{f,\varepsilon}$, by Proposition 10.2 (and in fact at a distance $O(\varepsilon^{\frac{1}{2}})$ from the spectrum of $\mathcal{B}_{f,\varepsilon}$) , so that

$$\|\partial_z g\|_\infty \leq O(\varepsilon^{-\frac{1}{2}} \varepsilon^{-12}) \left(O(\varepsilon^{-12}) \||(\mathcal{B}_{f,\varepsilon} - 1) g\|| \right)$$

hence $\||(\mathcal{B}_{f,\varepsilon} - 1)^{-1}\|| \leq O(\varepsilon^{-25})$. Now existence and uniqueness of a solution follow in exactly the same way as in the proof of Theorem 10.1 on a ball of radius ε^{26} in S_1 .

COROLLARY 10.10 .

The operator $\mathcal{A}_{\phi_\varepsilon}$ has spectrum as described in Proposition 10.2. In particular $\lambda_0 = 2$, $e_0 = \phi_\varepsilon$; $\lambda_1 = 2c^{-\frac{1}{2}}$, $e_1 = z\phi_\varepsilon$; $\lambda_3 = c^{\frac{1}{2}}$, $e_3 = \partial_z \phi_\varepsilon$.

Proof : The first part follows from the smallness of r_ε in Eq. (10.24). The second part follows from the fact that $\mathcal{N}_\varepsilon(\phi_\varepsilon) = \phi_\varepsilon$.

In Section 11, we shall need the following estimate.

THEOREM 10.11 . For sufficiently small $\varepsilon > 0$ one has for all σ, $0 < \sigma \leq \gamma$

(i) $\mathcal{A}_{\phi_\varepsilon} - 1$ is a bounded, invertible operator on $L_{2,\sigma}$. Its inverse is a norm continuous function of $\varepsilon > 0$, and it is bounded in norm by $O(\varepsilon^{-1})$.

(ii)   Let $\|g - \phi_\varepsilon\|_{4,\tau} = O(\varepsilon^{5/4})$. Then $(\mathcal{A}_{g,\varepsilon} - 1)^{-1}$ is a bounded operator of norm $O(\varepsilon^{-1})$ on $L_{2,\sigma}$ and it is norm continuous as a function of $g \in L_{4,\tau}$.

Proof : Since $\phi_\varepsilon \in L_{4,\tau}$ $\forall$ $\tau > 0$, we find by Lemma 10.3 that $\mathcal{A}_{\phi_\varepsilon}$ is compact from $L_{2,\sigma}$ to $L_{2,\sigma}$. For small $\varepsilon \geq 0$ we have $L_{2,\sigma} \subset L_{2,1-c_\varepsilon^{-1}}$. Therefore the spectrum of $\mathcal{A}_{\phi_\varepsilon} - 1$ consists of a point with multiplicity one near $O(\varepsilon)$ and a remainder bounded away from zero. The bound on $(\mathcal{A}_{\phi_\varepsilon} - 1)^{-1}$ is complicated by the fact that $\mathcal{A}_\varepsilon$ is not symmetric. Let P_ε be the orthogonal projection onto ϕ_ε in $L_{2,\sigma}$, $P_\varepsilon^\perp = 1 - P_\varepsilon$. Then one has

LEMMA 10.12.   The operator $P_\varepsilon^\perp(\mathcal{A}_{\phi_\varepsilon} - 1) P_\varepsilon^\perp$ is invertible on $P_\varepsilon^\perp L_{2,\sigma}$ and the norm of its inverse is uniformly bounded for $\varepsilon > 0$ sufficiently small.

We postpone the proof of this lemma and continue the proof of Theorem 10.11. Consider the "matrix"

$$
\mathcal{A}_{\phi_\varepsilon} - 1 = \begin{pmatrix} P_\varepsilon (\mathcal{A}_{\phi_\varepsilon} - 1) P_\varepsilon & P_\varepsilon (\mathcal{A}_{\phi_\varepsilon} - 1) P_\varepsilon^{\perp} \\ \\ 0 & P_\varepsilon^{\perp}(\mathcal{A}_{\phi_\varepsilon} - 1) P_\varepsilon^{\perp} \end{pmatrix} \tag{10.25}
$$

on

$$ P_\varepsilon L_{2,\sigma} \oplus P_\varepsilon^{\perp} L_{2,\sigma} \;. $$

The element $P_\varepsilon (\mathcal{A}_{\phi_\varepsilon} - 1) P_\varepsilon$ is invertible and its inverse is bounded by $O(\varepsilon^{-1})$ on $P_\varepsilon L_{2,\sigma}$, as a consequence of Corollary 10.10. The operator $P_\varepsilon (\mathcal{A}_{\phi_\varepsilon} - 1) P_\varepsilon^{\perp}$ is rank 1 on $L_{2,\sigma}$ and its norm is bounded as a function of small $\varepsilon \geq 0$ for fixed σ, as can be seen by explicitly calculating the Hilbert-Schmidt norm of $\mathcal{A}_{\phi_\varepsilon}$ on $L_{2,\sigma}$. Therefore the inverse of (10.25), which is

$$
\begin{pmatrix} (P_\varepsilon (\mathcal{A}_{\phi_\varepsilon} -1) P_\varepsilon)^{-1} & - (P_\varepsilon (\mathcal{A}_{\phi_\varepsilon} -1) P_\varepsilon)^{-1} (P_\varepsilon (\mathcal{A}_{\phi_\varepsilon} -1) P_\varepsilon^{\perp}) (P_\varepsilon^{\perp}(\mathcal{A}_{\phi_\varepsilon} -1) P_\varepsilon^{\perp})^{-1} \\ \\ 0 & (P_\varepsilon^{\perp}(\mathcal{A}_{\phi_\varepsilon} - 1) P_\varepsilon^{\perp})^{-1} \end{pmatrix} \tag{10.26}
$$

is bounded in norm by $O(\varepsilon^{-1})$, (the sum of the norms of the matrix elements). This proves (i), up to the norm continuity.

By Lemma 10.3, $\| \mathcal{A}_{g,\varepsilon} - \mathcal{A}_{\phi_\varepsilon,\varepsilon} \|_{2,\sigma} \leq O(\varepsilon^{5/4})$. The assertion follows now by (i) and standard perturbation theory [35, IV, Theorem 1.16]. This completes the proof of Theorem 10.11, (ii). By the continuity of ϕ_ε in $L_{4,\tau}$ the remainder of Theorem 10.11, (i) follows.

Proof of Lemma 10.12 : We first note that $\|\phi_\varepsilon - 1\|_{4,\tau} \leq O(\varepsilon)$, so

that by Lemma 10.3, $\|\mathcal{A}_{\phi_\varepsilon,\varepsilon} - \mathcal{A}_{1,0}\|_{2,\sigma} = O(\varepsilon)$. It suffices thus to show

(by [35, IV, Theorem 1.16]) that $P_\varepsilon^\perp (\mathcal{A}_{1,0} - 1) P_\varepsilon^\perp$ has a bounded inverse.

Similarly, we note that with $H_{4,\varepsilon}(x) = h_4(c(\varepsilon),x)$, cf Eq.(3.8),

$$\|e_4 - H_{4,0}\|_{2,\sigma} \leq \|e_4 - H_{4,\varepsilon}\|_{2,\sigma} + \|H_{4,\varepsilon} - H_{4,0}\|_{2,\sigma} < O(\varepsilon^{4/5}) + O(\varepsilon),$$

by the definition of Hermite polynomials. Therefore $\| P_\varepsilon^\perp - P_{H_{4,0}}^\perp \|_{2,\sigma} =$

$\| P_{H_{4,0}} - P_\varepsilon \|_{2,\sigma} = O(\varepsilon^{4/5})$ and it suffices to show the bounded inverti-

bility of $P^\perp (\mathcal{A}_{1,0} - 1) P^\perp$, where $P^\perp = 1 - P_{H_{4,0}}$ on $L_{2,\sigma}$. Now $\mathcal{A}_{1,0}$

is compact on $L_{2,\sigma}$, and hence 1 is at most an isolated eigenvalue

of $P^\perp \mathcal{A}_{1,0} P^\perp$. Suppose ψ is in the nullspace of $P^\perp (\mathcal{A}_{1,0} - 1) P^\perp$. Then

there is a λ such that $(\mathcal{A}_{1,0} - 1) P^\perp \psi = \lambda H_{4,0}$ Since $L_{2,\sigma} \subset L_{2,\gamma}$ this

equality holds on $L_{2,\gamma}$, so that (by the selfadjointness of $\mathcal{A}_{1,0}$ on

$L_{2,\gamma}$), $\lambda = 0$ and $P^\perp \psi = \lambda' H_{4,0}$. Going back to $L_{2,\sigma}$, we see that

$\lambda' = 0$. Hence $P^\perp (\mathcal{A}_{1,0} - 1) P^\perp$ is invertible on $P^\perp L_{2,\sigma}$ and has a

bounded inverse (for fixed σ). This proves Lemma 10.12.

PROPOSITION 10.12.   The function ϕ_ε is strictly positive.

Proof : Suppose ϕ_ε takes some strictly negative value. As ϕ_ε is

continuous and as it goes to zero at infinity it would have an absolute

negative minimum $-m_0 (m_0 > 0)$, and there would be a real number z_0 such

that $\phi_\varepsilon(z_0) = -m_0$. From the equation $\mathcal{N}(\phi_\varepsilon) = \phi_\varepsilon$ we have

$$\phi_\varepsilon(z_0) = \pi^{-\frac{1}{2}} \int \exp(-u^2) \, \phi_\varepsilon(z_0 c^{-\frac{1}{2}} - u) \, \phi_\varepsilon(z_0 c^{-\frac{1}{2}} + u) \, du$$

$$= 2\pi^{-\frac{1}{2}} \int_0^{+\infty} \exp(-u^2) \, \phi_\varepsilon(z_0 c^{-\frac{1}{2}} - u) \, \phi_\varepsilon(z_0 c^{-\frac{1}{2}} + u) \, du$$

$$= 2\pi^{-\frac{1}{2}} \int_0^{10} \exp(-u^2) \, \phi_\varepsilon(z_0 c^{-\frac{1}{2}} - u) \, \phi_\varepsilon(z_0 c^{-\frac{1}{2}} + u) \, du +$$

$$+ \ 2/\sqrt{\pi} \int_{10}^{+\infty} \exp(-u^2) \ \phi_\varepsilon \ (z_0 \ c^{-\frac{1}{2}} - u) \ \phi_\varepsilon \ (z_0 \ c^{-\frac{1}{2}} + u) \ du$$

$$\geq 2\pi^{-\frac{1}{2}} \int_0^{10} \exp(-u^2) \ \phi_\varepsilon \ (z_0 \ c^{-\frac{1}{2}} - u) \ \phi_\varepsilon \ (z_0 \ c^{-\frac{1}{2}} + u) \ du \ +$$

$$+ \ 2/\sqrt{\pi} \ \exp(-50) \ (-m_0) \ (1 + \varepsilon^{\frac{1}{4}}) \int_{10}^{+\infty} \exp(-u^2 / 2) \ du$$

$$\geq 2\pi^{-\frac{1}{2}} \int_0^{10} \exp(-u^2) \ \phi_\varepsilon \ (z_0 \ c^{-\frac{1}{2}} - u) \ \phi_\varepsilon \ (z_0 \ c^{-\frac{1}{2}} + u) \ du$$

$$- \ m_0 \ 2^{\frac{1}{2}} \ (1 + \varepsilon^{\frac{1}{4}}) \ \exp(-50)$$

where the bound $|\phi_\varepsilon| < (1 + \varepsilon^{\frac{1}{4}})$ comes from the representation

$$\phi_\varepsilon(x) \ = \ f_\varepsilon(x) \ + \ O(\varepsilon^{11}) \ \exp(-\varepsilon\theta/2 \ x^4) \quad .$$

From the same representation, it is easy to see that $|z_0| > 10^3 \ \varepsilon^{-\frac{1}{4}}$.
Now we have

$$\phi_\varepsilon(z_0 \ c^{-\frac{1}{2}} - u) \ \phi_\varepsilon \ (z_0 \ c^{-\frac{1}{2}} + u) \ \geq \ - \ m_0 \ \exp\left(-(\varepsilon\theta/4)(z_0 \ c^{-\frac{1}{2}} - 10)^4\right)$$

if $0 \leq u \leq 10$ and then

$$\phi_\varepsilon(z_0 \ c^{-\frac{1}{2}} - u) \ \phi_\varepsilon \ (z_0 \ c^{-\frac{1}{2}} + u) \ \geq \ - \ m_0 \ \exp(-\theta \ 10^8) \quad .$$

From the equation we have

$$- \ m_0 \ \geq \ 2\pi^{-\frac{1}{2}} \ 10 \ \left(-m_0 \ \exp(-\theta \ 10^8)\right) - m_0 \ (1 + \varepsilon^{\frac{1}{4}}) \ \exp(-50) \ 2^{\frac{1}{2}} \ \geq \ - \ m_0 \ /2$$

which is in contradiction with $m_0 > 0$. Thus ϕ_ε is nowhere negative,
and by applying once more $\mathcal{N}'$, we get the result.

Remarks on Section 10 :

The main result, the existence of ϕ_ε has been proved before by Bleher and Sinai in their fundamental paper [16], cf the "Remarks on Section 3".

The functional analytic apparatus we are using here can be found for questions of topology in

[34] DUNFORD-SCHWARTZ . Linear operators. Part I : General theory;
   Part II : Spectral theory. New York Interscience 1958, 1963,

while a good reference for the perturbation theory is

[35] T. KATO. Perturbation theory for linear operators.
   Berlin - Heidelberg - New York. Springer, 1966.

The hypercontractive estimates were first given by Glimm in a special case and later formulated and proved in full generality by Nelson in

[36] E. NELSON. The Free Markoff Field. J. Functional Anal. 12, 211-227
   (1973).

A nice proof which gives connections to Orlitz-Spaces has been given in

[37] L. GROSS. Logarithmic Sobolev Inequalities. Amer. J. Math. 97, 1061
   (1975).

The fact that the inequality follows from the ordinary Sobolev inequalities has been shown by Sénéor (private communication), by using the bounds given by

[38] T. AUBIN. Problèmes isopérimétriques et espaces de Sobolev.
   C.R. Acad. Sc. Paris 280, A 279 (1975).

A very elegant new proof can be found in

[39] H.J. BRASCAMP, E.H. LIEB. Best constants in Young's inequality, its converse and its generalization to more than three functions. Adv. Math. 20, 151 (1976).

11. Differentiability of ϕ_ϵ

It follows from the construction of ϕ_ϵ that it has an asymptotic expansion in powers of ϵ in L_∞ (i.e. ϕ_ϵ equals $\exp(-\epsilon\theta x^4)P_K(\epsilon,x) + O(\epsilon^{K/2})\exp(-\epsilon\theta x^4/2)$, where P_K is the correct polynomial up to order K in ϵ). In this section, we do better by proving the

THEOREM 11.1. For $\epsilon \in [0,\epsilon_0)$, the function $\epsilon \to \phi_\epsilon$ is C^∞ in $L_{2,\gamma}$. The function $\epsilon \to \tilde{\phi}_\epsilon$ extends to a real analytic function on $(0,\epsilon_0)$ with values in $E_\rho = \{f\,|\,f$ entire, $|f(z)| \le \exp(\rho\,|z|^2)\}$, $\rho > 0$. ($\tilde{\phi}_\epsilon$ is the Fourier transform of ϕ_ϵ). Hence the first few eigenvalues of $\mathcal{A}_{\phi_\epsilon}$ are real analytic functions on $(0,\epsilon_0)$.

THEOREM 11.2. The function ϕ_ϵ is entire in z and satisfies a bound

$$|\phi_\epsilon(z)| \le L\,\exp(A\,|{\rm Im}z|^{4+0(\epsilon)})\,.$$

In fact it is in the class $\mathcal{S}_{3/4+0(\epsilon)}^{1/4-0(\epsilon)}$ of Gelfand-Shilov [19].

The harder part of the proofs of these statements has already been given in our paper [18]; we repeat it here to make the Lecture Notes self-contained. It is typical for the kind of results stated above to follow from the linear properties of the model, i.e. from a study of $\mathcal{A}_{\phi_\epsilon,\epsilon}$.

PROPOSITION 11.3. For sufficiently small $\epsilon > 0$, ϕ_ϵ is a C^∞ function of z as an element of $L_{2,\sigma}$, for all $\sigma > 0$.

Proof : We show inductively that ϕ_ϵ is C^N. The case $N = 0,1$ follows from Theorem 10.1 and Corollary 10.9, since for all σ, $L_\infty \subset L_{2,\sigma}$. We suppose the result holds for $f_j = \partial_z^j\phi_\epsilon$, $j = 0,\dots$, $N-1$. By the equation $\mathcal{N}_\epsilon(\phi_\epsilon) = \phi_\epsilon$, we have with $c = c_\epsilon$,

$$f_{N-1}(z) = c^{-(N-1)/2} \pi^{-\frac{1}{2}} \sum_{j=0}^{N-1} \binom{N-1}{j} \int e^{-u^2} f_j(zc^{-\frac{1}{2}}+u) f_{N-1-j}(zc^{-\frac{1}{2}}-u) du. \quad (A_{N-1})$$

Since $c \sim 2^{\frac{1}{2}}$ this equality holds on $L_{2,\sigma/9\cdot3/c} \subset L_{2,\sigma/3}$, by Eq. (10.9).
(This relation implies $f_j \in L_{2,\sigma/9}$ for $j = 1,\ldots N-1$.) Define also

$$g_N = c^{-N/2} \pi^{-\frac{1}{2}} \sum_{k=1}^{N-1} \binom{N}{k} \int e^{-u^2} f_k(zc^{-\frac{1}{2}}+u) f_{N-k}(zc^{-\frac{1}{2}}-u) du$$

$$+ 2c^{-(N-1)/2} \pi^{-\frac{1}{2}} \int e^{-u^2} u\,du\,(f_{N-1}(zc^{-\frac{1}{2}}+u) f_0(zc^{-\frac{1}{2}}-u)$$

$$+ f_0(zc^{-\frac{1}{2}}+u) f_{N-1}(zc^{-\frac{1}{2}}-u))$$

$$- c^{-N/2} \pi^{-\frac{1}{2}} \int e^{-u^2} (f_{N-1}(zc^{-\frac{1}{2}}+u) f_1(zc^{-\frac{1}{2}}-u)$$

$$+ f_1(zc^{-\frac{1}{2}}+u) f_{N-1}(zc^{-\frac{1}{2}}-u)) du. \quad (B_N)$$

By the inductive assumption and by (10.9), g_N is defined on
$L_{2,\sigma/9\cdot3/c} \subset L_{2,\sigma/3}$ and bounded uniformly in $0 \le \epsilon$. Using an integration by parts formula, it is easy to see that g_N is a candidate for
$f_N = \partial_z^N \phi_\epsilon$. By the inductive assumption and partial integration, g_N is
the derivative of g_{N-1} with respect to z on $L_{2,\sigma/3}$. Also $g_{N-1} = f_{N-1}$
on $L_{2,\sigma/3}$, since the corresponding r.h.s. of A_{N-1} and B_{N-1} coincide on
this space. Therefore $g_N = \partial_z g_{N-1} = \partial_z f_{N-1}$, i.e. f_{N-1} is differentiable
on $L_{2,\sigma/3}$ and in fact continuously differentiable as can be seen by a
change of variables $u \to u \mp zc^{-\frac{1}{2}}$; its derivative is then equal to the
r.h.s. of A_N, as an element of $L_{2,\sigma} \supset L_{2,\sigma/9\cdot3/c}$. The induction step is
complete.

Proof of Theorem 11.2:

From Eq. A_N it is easy to see that for real z the inductive bound

$|\partial_z^j \phi_\varepsilon(z)| \leq j! \ c^{j+1}$ holds. Hence ϕ_ε is the restriction to the real axis of a function (called again ϕ_ε) which is analytic in a strip about the real axis and which is a solution of the equation in the strip. Each time we substitute this solution in the identity $\mathcal{N}_\varepsilon(\phi_\varepsilon) = \phi_\varepsilon$ the strip is enlarged by a factor $c^{\frac{1}{2}}$. Hence we get the fact that ϕ_ε is entire, and also the asserted bound on its increase at infinity.

Proof of Theorem 11.1:

We proceed in several steps. We first show in Lemma 11.4 that ϕ_ε is C^N for $\varepsilon > 0$ sufficiently small. Then we show in Lemma 11.5 that $\partial_z^k \phi_\varepsilon(z)$ is bounded as $\varepsilon \to 0$. We finally deduce the differentiability of ϕ_ε at $\varepsilon = 0$.

LEMMA 11.4. For all $N \geq 0$, $\sigma > 0$ there is an $\varepsilon_2 > 0$ such that for $0 < \varepsilon < \varepsilon_2$ the function $\phi_\varepsilon(z)$ is C^N in ε and z as an element of $M_{2,\sigma,\varepsilon_2}$ [1]).

Proof. As in the proof of Proposition 11.3 we shall show recursively the following properties.

P'_N : For $k = 0,1,2,\ldots,$ $\partial_z^k \partial_\varepsilon^N \phi_\varepsilon$ is in $L_{2,\sigma}$ for $0 < \varepsilon$ and it is continuous in ε.

P_N : $(\partial_\varepsilon^N \phi_\varepsilon)(z)$

$$= \pi^{-\frac{1}{2}} \sum_{j=0}^{N} \int e^{-u^2} \binom{N}{j} \partial_\varepsilon^j (\phi_\varepsilon(zc^{-\frac{1}{2}}+u)) \partial_\varepsilon^{N-j}(\phi_\varepsilon(zc_\varepsilon^{-\frac{1}{2}}-u)) du.$$

Note that P'_0 is a trivial consequence of Proposition 11.3 and Lemma 10.3. Also P_0 expresses the fact that $\phi_\varepsilon \in L_{2,\sigma/3}$ solves $\mathcal{N}_\varepsilon(\phi_\varepsilon) = \phi_\varepsilon$.

---

[1]) The topology of $M_{2,\sigma,\varepsilon_0}$ is given through the norm $\sup\limits_{0 \leq \varepsilon < \varepsilon_0} \|\phi_\varepsilon\|_{2,\sigma}$, where $\|\cdot\|_{2,\sigma}$ is the norm of $L_{2,\sigma}$.

Suppose now that P_j, P_j' hold for $j \leq N$. In particular, we have on $L_{2,\sigma}$,

$$(\partial_\varepsilon^N \phi_\varepsilon)(z) = \pi^{-\frac{1}{2}} \sum_{j=1}^{N-1} \int e^{-u^2} \binom{N}{j} \partial_\varepsilon^j (\phi_\varepsilon(zc_\varepsilon^{-\frac{1}{2}} + u)) \cdot \partial_\varepsilon^{N-j}(\phi_\varepsilon(zc_\varepsilon^{-\frac{1}{2}} - u)) du$$

$$+ 2\pi^{-\frac{1}{2}} \int e^{-u^2} \partial_\varepsilon^N (\phi_\varepsilon(zc_\varepsilon^{-\frac{1}{2}} + u)) \phi_\varepsilon(zc_\varepsilon^{-\frac{1}{2}} - u) du \qquad (11.1)$$

$$= g_\varepsilon^{(N)}(z) + 2\pi^{-\frac{1}{2}} \int e^{-u^2} \phi_\varepsilon(zc_\varepsilon^{-\frac{1}{2}} + u)(\partial_\varepsilon^N \phi_\varepsilon)(zc_\varepsilon^{-\frac{1}{2}} - u) du.$$

By the chain rule, we find

$$g_\varepsilon^{(N)}(z) = \pi^{-\frac{1}{2}} \sum_{j=1}^{N-1} \int e^{-u^2} \binom{N}{j} \partial_\varepsilon^j (\phi_\varepsilon(zc_\varepsilon^{-\frac{1}{2}} + u)) \partial_\varepsilon^{N-j}(\phi_\varepsilon(zc_\varepsilon^{-\frac{1}{2}} - u)) du$$

$$+ 2\pi^{-\frac{1}{2}} \sum_{\substack{j+\Sigma \ell n_\ell = N \\ \ell \geq 1, \, j < N}} \frac{N!}{j! \, \pi_\ell \, n_\ell! \ell!^{n_\ell}} \prod_\ell \{z \partial_\varepsilon^\ell c_\varepsilon^{-\frac{1}{2}}\}^{n_\ell}$$

$$\cdot \int e^{-u^2} (\partial_\varepsilon^j \partial_z^{\Sigma n_\ell} \phi_\varepsilon)(zc_\varepsilon^{-\frac{1}{2}} + u) \phi_\varepsilon(zc_\varepsilon^{-\frac{1}{2}} - u) du.$$

$$(11.2)$$

These expressions are well defined on $L_{2,\sigma}$ by the inductive assumption P_N. We can now form for $0 < \varepsilon$, ε' on $L_{2,\sigma}$,

$$(\partial_\varepsilon^N \phi_\varepsilon)(z) - (\partial_\varepsilon^N, \phi_{\varepsilon'})(z) = g_\varepsilon^{(N)}(z) - g_{\varepsilon'}^{(N)}(z)$$

$$+ 2\pi^{-\frac{1}{2}} \int e^{-u^2} \phi_\varepsilon(zc_\varepsilon^{-\frac{1}{2}} + u)[(\partial_\varepsilon^N \phi_\varepsilon)(zc_\varepsilon^{-\frac{1}{2}} - u) - (\partial_\varepsilon^N, \phi_{\varepsilon'})(zc_\varepsilon^{-\frac{1}{2}} - u)] du$$

$$+ 2\pi^{-\frac{1}{2}} \int e^{-u^2} \phi_\varepsilon(zc_\varepsilon^{-\frac{1}{2}} + u)[(\partial_\varepsilon^N, \phi_{\varepsilon'})(zc_\varepsilon^{-\frac{1}{2}} - u) - (\partial_\varepsilon^N, \phi_{\varepsilon'})(zc_{\varepsilon'}^{-\frac{1}{2}} - u)] du$$

$$+ 2\pi^{-\frac{1}{2}} \int e^{-u^2} [\phi_\varepsilon(zc_\varepsilon^{-\frac{1}{2}} + u) - \phi_{\varepsilon'}(zc_{\varepsilon'}^{-\frac{1}{2}} + u)](\partial_\varepsilon^N, \phi_{\varepsilon'})(zc_{\varepsilon'}^{-\frac{1}{2}} - u) du$$

$$= g_\varepsilon^{(N)}(z) - g_{\varepsilon'}^{(N)}(z) + \sum_{k=1}^{3} g_k^{(N)}(\varepsilon, \varepsilon', z).$$

Solving for $(\partial_\varepsilon^N \phi_\varepsilon)(z) - (\partial_\varepsilon^N, \phi_{\varepsilon'})(z)$, we get

$$(1 - \mathcal{A}_{\phi_\epsilon, \epsilon})(\partial_\epsilon^N \phi_\epsilon - \partial_{\epsilon'}^N \phi_{\epsilon'}) = g_\epsilon^{(N)} - g_{\epsilon'}^{(N)} + \sum_{k=2,3} g_k^{(N)}(\epsilon, \epsilon', \cdot), \qquad (11.3)$$

and this is well defined on $L_{2,\sigma}$. Since $\partial_\epsilon^N \phi_\epsilon$ and $\partial_{\epsilon'}^N \phi_{\epsilon'}$ are continuously differentiable as functions of z, by P_N', we may rewrite $g_2^{(N)} + g_3^{(N)}$ as "derivatives plus remainder", i.e.

$$g_2^{(N)}(\epsilon, \epsilon', z) = (\epsilon - \epsilon') \cdot 2\pi^{-\frac{1}{2}} z \partial_\epsilon c_\epsilon^{-\frac{1}{2}} \int e^{-u^2} \phi_\epsilon (zc_\epsilon^{-\frac{1}{2}} + u)$$

$$(\partial_z \partial_\epsilon^N \phi_\epsilon)(zc_\epsilon^{-\frac{1}{2}} - u) du + O((\epsilon - \epsilon')^2), \qquad (11.4)$$

on $L_{2,\sigma}$ provided $0 < \epsilon, \epsilon'$.

We similarly have

$$g_3^{(N)}(\epsilon, \epsilon', z) = (\epsilon - \epsilon') \cdot 2\pi^{-\frac{1}{2}} \int e^{-u^2} (\partial_\epsilon \phi_\epsilon)(zc_\epsilon^{-\frac{1}{2}} + u)(\partial_{\epsilon'}^N \phi_{\epsilon'})(zc_{\epsilon'}^{-\frac{1}{2}} - u) du$$

$$+ (\epsilon - \epsilon') \cdot 2\pi^{-\frac{1}{2}} z \partial_\epsilon c_\epsilon^{-\frac{1}{2}} \int e^{-u^2} (\partial_z \phi_\epsilon)(zc_\epsilon^{-\frac{1}{2}} + u)$$

$$\cdot (\partial_{\epsilon'}^N \phi_{\epsilon'})(zc_{\epsilon'}^{-\frac{1}{2}} - u) du + O((\epsilon - \epsilon')^2), \qquad (11.5)$$

under the same conditions as before, if $N \geq 1$. For $N = 0$, Eq. (11.3) is replaced by

$$[(1 - \tfrac{1}{2} \mathcal{A}_{\phi_\epsilon, \epsilon} - \tfrac{1}{2} \mathcal{A}_{\phi_{\epsilon'}, \epsilon'})(\phi_\epsilon - \phi_{\epsilon'})](z)$$

$$\qquad (11.3)'$$

$$= \pi^{-\frac{1}{2}} \int e^{-u^2} \{\phi_{\epsilon'}(zc_\epsilon^{-\frac{1}{2}} + u)\phi_{\epsilon'}(zc_\epsilon^{-\frac{1}{2}} - u) - \phi_{\epsilon'}(zc_{\epsilon'}^{-\frac{1}{2}} + u)\phi_{\epsilon'}(zc_{\epsilon'}^{-\frac{1}{2}} - u)\} du$$

$$= 2\pi^{-\frac{1}{2}}(\epsilon - \epsilon') \int e^{-u^2} \{\partial_z \phi_\epsilon (zc_\epsilon^{-\frac{1}{2}} + u)\phi_\epsilon (zc_\epsilon^{-\frac{1}{2}} - u) z \partial_\epsilon c_\epsilon^{-\frac{1}{2}}\} du + O((\epsilon - \epsilon')^2).$$

Henceforth we only discuss Eq. (11.3), the case (11.3)' is analogous. By Theorem 10.11, $\mathcal{A}_{\phi_\epsilon, \epsilon} - 1 = \mathcal{A}_\epsilon - 1$ is invertible on $L_{2,\sigma}$ for sufficiently small $\epsilon > 0$, and its inverse is bounded in norm by $O(\epsilon^{-1})$, and continuous in $\varepsilon > 0$. Therefore

$$\partial_\varepsilon (\partial_\varepsilon^N \phi_\varepsilon) = \lim_{\varepsilon' \to \varepsilon} (\varepsilon - \varepsilon')^{-1} (\partial_\varepsilon^N \phi_\varepsilon - \partial_{\varepsilon'}^N \phi_{\varepsilon'}) \qquad (11.6)$$

$$= (1 - \mathcal{A}_\varepsilon)^{-1} (\partial_\varepsilon g_\varepsilon^{(N)} + \sum_{k=2,3} \partial_\varepsilon \cdot g_k (\varepsilon, \varepsilon', \cdot)) \Big|_{\varepsilon'=\varepsilon}$$

is continuous in $\varepsilon > 0$, on $L_{2,\sigma}$.

Multiplying (11.6) by $1 - \mathcal{A}_\varepsilon$ on both sides, one gets the relation P'_{N+1}. Next we show that $\partial_\varepsilon^{N+1} \phi_\varepsilon$ is differentiable in z. Using the relation P'_{N+1} and the inductive assumption P_N it is clear from Eq. (11.1) that it suffices to show the differentiability of

$$\int e^{-u^2} (\partial_\varepsilon^{N+1} \phi_\varepsilon) (zc_\varepsilon^{-\frac{1}{2}} + u) \phi_\varepsilon (zc_\varepsilon^{-\frac{1}{2}} - u) \, du.$$

But this equals

$$\int e^{-(u - zc_\varepsilon^{-\frac{1}{2}})^2} \partial_\varepsilon^{N+1} \phi_\varepsilon (u) \phi_\varepsilon (2zc_\varepsilon^{-\frac{1}{2}} - u) \, du, \qquad (11.7)$$

and the assertion follows now by Theorem 11.2 and by the bound P'_{N+1} on $\partial_\varepsilon^{N+1} \phi_\varepsilon$.

We now work towards differentiability at $\varepsilon = 0$. Our first result is

LEMMA 11.5. For all $\sigma > 0$ and $K \in \mathbb{Z}^+$, the function $\partial_z^K \phi_\varepsilon$ is uniformly bounded in $0 \le \varepsilon$, as an element of $L_{2,\sigma}$.

Proof : This follows immediately from Theorem 11.2 and the fact that the assertion is true for $K = 0,1$, by Theorem 10.8.

We have already seen from the proof of Theorem 10.8 that ϕ_ε has an asymptotic expansion of the form $\exp(-\varepsilon \theta x^4) P(\varepsilon, x)$ on L_∞, and hence it has an asymptotic expansion of the form

$$\phi_\varepsilon = Q_k (\varepsilon, \cdot) + O(\varepsilon^{k+1}) \qquad (11.8)$$

on all $L_{2,\sigma}$ spaces, where $Q_k(\varepsilon,x)$ is a polynomial of degree k in ε and 4k in x.

LEMMA 11.6. One has for sufficiently small $\varepsilon \geq 0$ for all k, $n \geq 0$ and $\sigma > 0$ the representation

$$\partial_z^n \phi_\varepsilon = \partial_z^n P_{k+3}(\varepsilon,\cdot) + R'_{k,\varepsilon,n} \quad , \tag{11.9}$$

with

$$\| R'_{k,\varepsilon,n} \|_{2,\sigma} \leq \varepsilon^k C(n,k,\sigma) .$$

Proof : The case $n = 0$ is covered in Eq. (11.8). To prove the case $n = 1$, we write first Eq. (11.8) for $k + 1$,

$$\phi_\varepsilon = Q_k(\varepsilon,\cdot) + \varepsilon^k R'_\varepsilon \quad , \tag{11.10}$$

with $\| R'_\varepsilon \|_{2,\sigma/3} = O(\varepsilon)$, for sufficiently small $\varepsilon \geq 0$. Since ϕ_ε and $P_{k+3}(\varepsilon,\cdot)$ are differentiable in z on $L_{2,\sigma/3}$ we find

$$\partial_z \phi_\varepsilon = \partial_z P_{k+3}(\varepsilon,\cdot) + \varepsilon^k \partial_z R'_\varepsilon . \tag{11.11}$$

Using now

$$\partial_z \phi_\varepsilon(z) = 2\pi^{-\frac{1}{2}} c_\varepsilon^{-\frac{1}{2}} \int e^{-u^2} \phi_\varepsilon(zc_\varepsilon^{-\frac{1}{2}} + u) \partial_z \phi_\varepsilon(zc_\varepsilon^{-\frac{1}{2}} - u) du, \tag{11.12}$$

we find on $L_{2,\sigma}$ the identity

$$\partial_z P_{k+3}(\varepsilon,z) + \varepsilon^k \partial_z R'_\varepsilon(z)$$

$$= 2\pi^{-\frac{1}{2}} c_\varepsilon^{-\frac{1}{2}} \int e^{-u^2} P_{k+3}(\varepsilon, zc_\varepsilon^{-\frac{1}{2}} + u)(\partial_z \phi_\varepsilon)(zc_\varepsilon^{-\frac{1}{2}} - u) du$$

$$+ 2\pi^{-\frac{1}{2}} c_\varepsilon^{-\frac{1}{2}} \int e^{-u^2} \varepsilon^k R'_\varepsilon(zc_\varepsilon^{-\frac{1}{2}} + u)(\partial_z \phi_\varepsilon)(zc_\varepsilon^{-\frac{1}{2}} - u) du,$$

which becomes upon integrating by parts

$$\varepsilon^k \partial_z R'_\varepsilon(z) = -\partial_z P_{k+3}(\varepsilon, z)$$

$$+ 2\pi^{-\frac{1}{2}} c_\varepsilon^{-\frac{1}{2}} \int e^{-u^2} \{ (\partial_z P_{k+3})(\varepsilon, zc_\varepsilon^{-\frac{1}{2}} + u) + 2u\, P_{k+3}(\varepsilon, zc_\varepsilon^{-\frac{1}{2}} + u) \} \phi_\varepsilon (zc_\varepsilon^{-\frac{1}{2}} - u)\, du$$

$$+ 2\pi^{-\frac{1}{2}} c_\varepsilon^{-\frac{1}{2}} \int e^{-u^2} \varepsilon^k R'_\varepsilon (zc_\varepsilon^{-\frac{1}{2}} + u)(\partial_z \phi_\varepsilon)(zc_\varepsilon^{-\frac{1}{2}} - u)\, du. \qquad (11.13)$$

By Eq. (11.8), Lemma 10.3 and Theorem 11.2, the second integral is $O(\varepsilon^{k+1})$ in $L_{2,\sigma}$. In the first integral of (11.13), we split ϕ_ε according to (11.10) and the term coming from R'_ε is bounded by $O(\varepsilon^{k+1})$ in $L_{2,\sigma}$. The other term is equal to

$$2\pi^{-\frac{1}{2}} c_\varepsilon^{-\frac{1}{2}} \int e^{-u^2} \{ (\partial_z P_{k+3})(\varepsilon, zc_\varepsilon^{-\frac{1}{2}} + u) + 2u\, P_{k+3}(\varepsilon, zc_\varepsilon^{-\frac{1}{2}} + u) \}$$

$$\cdot\ P_{k+3}(\varepsilon, zc_\varepsilon^{-\frac{1}{2}} - u)\, du \qquad (11.14)$$

$$= 2\pi^{-\frac{1}{2}} c_\varepsilon^{-\frac{1}{2}} \int e^{-u^2} P_{k+3}(\varepsilon, zc_\varepsilon^{-\frac{1}{2}} + u)(\partial_z P_{k+3})(\varepsilon, zc_\varepsilon^{-\frac{1}{2}} - u)\, du\ .$$

By perturbation theory, (11.14) is equal to $\partial_z P_{k+3}(\varepsilon, z)$ up to a polynomial which is of order $k+3$ in ε. Going back to (11.13) we see that $\| \partial_z R'_\varepsilon \|_{2,\sigma} = O(\varepsilon)$ so that it is in particular uniformly bounded in $\varepsilon \geq 0$, and well defined for $\varepsilon = 0$. This proves (11.9) for $n = 1$. The cases $n \geq 2$ follow by induction as in Theorem 11.2.

End of proof of Theorem 11.1. We show that for all k, N, ℓ, $k \geq N+3$, and $\sigma > 0$ one has for $\varepsilon \geq 0$ sufficiently small, depending on k, N, ℓ, σ, the representation

$$\partial_z^\ell \partial_\varepsilon^N \phi_\varepsilon = \partial_z^\ell \partial_\varepsilon^N P_k(\varepsilon, \cdot) + O(\varepsilon^{k-N-3}), \qquad (11.15)$$

on $L_{2,\sigma}$. This obviously implies Theorem 11.1. We prove (11.15) by induction on N. For $N = 0$ it is the content of Lemma 11.6. Suppose now (11.15) is true for $N \leq n$ on $L_{2,\sigma/3}$. By the property P'_{n+1} of Lemma 11.4, we have for $\varepsilon \geq 0$ on $L_{2,\sigma/3}$ the identity

$$(1 - \mathcal{A}_{\phi_\varepsilon, \varepsilon}) \, (\partial_\varepsilon^{n+1} \phi_\varepsilon) \, (z)$$

$$= \pi^{-\frac{1}{2}} \sum_{j=1}^{n} \int e^{-u^2} \binom{n+1}{j} (\partial_\varepsilon^j \phi_\varepsilon) (zc_\varepsilon^{-\frac{1}{2}} + u) (\partial_\varepsilon^{n+1-j} \phi_\varepsilon) (zc_\varepsilon^{-\frac{1}{2}} - u) \, du, \qquad (11.16)$$

so that by the induction hypothesis and the definition of P_k we have on $L_{2,\sigma/3}$,

$$(1 - \mathcal{A}_\varepsilon) \, (\partial_\varepsilon^{n+1} \phi_\varepsilon) = (1 - \mathcal{A}_\varepsilon) \, \partial_\varepsilon^{n+1} P_k (\varepsilon, \cdot) + O(\varepsilon^{k-n-3}). \qquad (11.17)$$

Applying Theorem 10.11, we get (11.15) for $N = n + 1$ and $\ell = 0$, and therefore the derivatives with respect to ε of ϕ_ε are bounded and can be extended to $\varepsilon = 0$. By the induction hypothesis, the terms on the r.h.s. of (11.16) are ℓ times differentiable in z so that $(1 - \mathcal{A}_\varepsilon) \partial_z^\ell \, (\partial_\varepsilon^{n+1} \phi_\varepsilon) (z)$ can be defined as a suitable sum of $\int e^{-u^2}$ times derivatives of the form

$$\partial_z^{\ell'} \partial_\varepsilon^j \phi_\varepsilon \;, \quad j = 0, \ldots n; \;\; \ell' = 0, \ldots \ell \;;$$

$$\partial_\varepsilon^p c_\varepsilon^{-\frac{1}{2}} \;, \quad u,$$

(apply the Eqs. (11.1), (11.2) to $\partial_\varepsilon^{n+1}$, solve for $(\partial_\varepsilon^{n+1} \phi_\varepsilon)$, differentiate both sides ℓ times with respect to z). We can now use the induction hypothesis (11.15) for $\ell' \le \ell$, $N \le n$ on $L_{2,\sigma/3}$ and then (11.15) follows for $\ell' \le \ell$, $N \le n + 1$ on $L_{2,\sigma}$ since no further powers of ε are lost by differentiating with respect to z. This completes the induction proof of (11.15).

It remains to prove the real analyticity. This will not be done here, because the proof is lengthy. It will be given in the Ph.D. thesis of P. Collet. The proof is a painful sequence of imbeddings and estimates of the Fourier transform of the fixed point equation, working mostly in the space of entire functions in z of order 2 (growth $\exp(\rho|z|^2)$).

12.   The Normal Form of the Flow

In our paper [18], we discussed already the normal form of the flow. Our proof was however wrong and should be disregarded. We thank D. Chillingworth and L. Guimaraez for pointing out the mistake.

The normal form of maps around a fixed point is widely discussed in the literature. There are two main lines of questions that can be asked:

1) Whether there is an invertible coordinate transformation which lets the map appear identical to its tangent map (at the fixed point).

2) Whether this coordinate transformation is differentiable.

For our purposes we need C^1, in order to identify the temperature in the computations of critical indices.

Both of the above goals fail partially in the case at hand. The problems are that $\mathcal{A}_{\phi_\varepsilon, \varepsilon}$ is not known to be invertible and that it has an eigenvalue very near to 1 (namely λ_4), and this makes a verification of the Sternberg conditions very hard, since even to prove only C^1, one may need to go to very high values of $\sum \alpha_j$ in Eq. (4.4), (namely up to order $O(\varepsilon^{-1})$). A further problem is that the spectrum of $\mathcal{A}_{\phi_\varepsilon, \varepsilon}$ near 0 is not well known at all. Our way out is to be more modest, and to linearize the transformation

$$T(\psi) \quad = \quad \mathcal{N}_\varepsilon (\phi_\varepsilon + \psi) \quad - \quad \phi_\varepsilon \quad ,$$

only in the most unstable direction. This is in fact sufficient for our physics analysis except for the case of the critical index of the magnetization, for which one needs a linearization in the two most unstable directions. This could be achieved by working out the Sternberg conditions to order $O(\varepsilon^2)$, and by pushing the analysis of this section

a bit further, but we do not work out the details in these Lecture

Notes.

We have already seen that $\mathcal{A}_{\phi_\varepsilon,\varepsilon}$ for $\varepsilon > 0$ is compact on L_∞ and has

spectrum described by Proposition 10.2.

Our next problem is to make sure that the part of the spectrum of

$\mathcal{A}_\varepsilon = \mathcal{A}_{\phi_\varepsilon,\varepsilon}$ corresponding to the eigenvalues of modulus less than 1

is a contraction. This is shown in the following theorem of Mather.

LEMMA 12.1.[25] <u>Let A be a linear continuous map from a Banach space</u>

<u>E onto itself, with a finite number of eigenvalues of modulus above 1</u>

<u>and the remainder of the spectrum in a circle of radius strictly less</u>

<u>than 1 around the origin. Then there is a spectral decomposition</u>

$E = E_1 \oplus E_2$, <u>invariant under A, and there are norms on E_1, E_2 equi-</u>

<u>valent to the original ones such that</u> $A\big|_{E_2}$ <u>is a contraction and</u>

$(A\big|_{E_1})^{-1}$ <u>is a contraction.</u>

(The norm is defined as follows: Let ρ be the radius of the circle,

$\rho' = (\rho+1)/2$, $A_2 = A\big|_{E_2}$. One has for sufficiently large n the inequa-

lity $\|A_2^n\|^{1/n} < \rho'$. Therefore $\|A_2^n\| < C \cdot (\rho')^n$ for some C and all n. Let

p be such that $C \cdot (\rho')^p < 1$. Then the new norm is defined by

$$\||f\|| = \sum_{q=0}^{p} \|A_2^q f\| ,$$

for $f \in E_2$.) In fact it follows from the proof of Theorem 10.1 that

$$\|f\|_\infty \leq \||f\|| \leq O(\varepsilon^{-13})\|f\|_\infty . \tag{12.1}$$

Let E_1 be the unstable subspace of $\mathcal{A}_{\phi_\varepsilon,\varepsilon}$ and E_2 the stable spectral subspace. Using the argument of Mather, there are equivalent norms on E_1 and E_2 such that

$$\| A_1 \|_{E_1} \quad = \quad 2 \, ,$$

$$\| A_1^{-1} \|_{E_1} \quad = \quad \lambda_2^{-1} \, ,$$

$$\| A_2 \|_{E_2} \quad = \quad \lambda_4 + \zeta \, ,$$

with $\zeta > 0$ but as small as we want and $A_i = P_{E_i} \cdot \mathcal{A}_{\phi_\varepsilon,\varepsilon}$. In the decomposition $E = L_\infty = E_1 + E_2$, T is of the form

$$T \begin{bmatrix} x \\ y \end{bmatrix} = \begin{bmatrix} A_1 x + N_1'(x,y) \\ A_2 y + N_2'(x,y) \end{bmatrix} \quad , \quad x \in E_1 \, , \, y \in E_2 \quad .$$

It is easily verified that $\mathcal{A}_{\phi_\varepsilon,\varepsilon}$ is 1-hyperbolic in the sense of [25] and hence it follows from [25, Corollary 5.4] that there exist locally stable and unstable manifolds $\mathcal{W}_s$ and $\mathcal{W}_u$, respectively, for T. These manifolds are the graphs of two C^∞ functions $f_s : E_2 \to E_1$ and

$f_u : E_1 \to E_2$. Let S be the C^∞ diffeomorphism given by

$$S^{-1} \begin{bmatrix} x \\ y \end{bmatrix} = \begin{bmatrix} x + f_s(y) \\ y + f_u(x) \end{bmatrix} \quad .$$

Then S is the map whose existence was asserted in Theorem 4.2. The transformation $t = S T S^{-1}$ has now E_1 and E_2 as its unstable and stable manifolds, respectively. It is of the form

$$t \begin{bmatrix} x \\ y \end{bmatrix} = \begin{bmatrix} x + M_1(x,y) \\ y + M_2(x,y) \end{bmatrix} \quad ,$$

with

$$M_1(0,y) = 0 \quad , \quad M_2(x,0) = 0 \quad .$$

The maps M_i, i= 1,2 are C^∞ and of "order two". Since there is no Sternberg relation between the eigenvalues which are of modulus greater than one, by Theorem 4.3, the map $t \big|_{E_1}$ can be linearized through a transformation which is C^k for sufficiently small $\varepsilon > 0$. We shall therefore assume without loss of generality in the sequel that $t \big|_{E_1} = A_1$.
We are going to isolate now the most dilating part of A_1. Let λ be the largest eigenvalue of A_1, and denote by e the associated eigenvector, while E_1' will denote the spectral supplement of e in E_1 , and $A_1' = P_{E_1'} A_1$.

THEOREM 12.2. There exists a C^1 diffeomorphism U, whose derivative is $O(\varepsilon)$-Hölder continuous such that U(0) = 0 and $t' = U t U^{-1}$ satisfies

$$t' \begin{bmatrix} x_o \\ x \\ y \end{bmatrix} = \begin{bmatrix} \lambda x_o \\ A_1' x + M(x_o, x, y) \\ A_2 y + M'(x_o, x, y) \end{bmatrix} \quad , \ x_o \in \mathbb{C}, \ x \in E_1', \ y \in E_2 \quad .$$

The functions M and M' are C^1 and will be given in the proof.

Proof: We shall look for a U of the form

$$U \begin{bmatrix} x_o \\ x \\ y \end{bmatrix} = \begin{bmatrix} x_o + R(x_o,x,y) \\ x \\ y \end{bmatrix} .$$

For this form, we obtain easily the equation for R. It is

$$\lambda R(x_o,x,y) = F_o(x_o,x,y) \tag{12.2}$$

$$+ R(\lambda x_o + F_o(x_o,x,y), \; A_1 x + F_1(x_o,x,y), \; A_2 y + F_2(x_o,x,y))$$

where $F_o = P_e M_1$, $F_1 = P_{E_1'} M_1$, $F_2 = M_2$. Due to the linearization of the unstable manifold, we have

$$F_2(x_o,x,0) = F_1(0,0,y) = F_o(0,0,y) = F_1(x_o,x,0) = F_o(x_o,x,0) = 0.$$

Therefore we can write

$$F_i(x_o,x,y) = L_i(x_o,x,y)y , \quad i = 0,1,2,$$

where L_i is a C^k function from E to the Banach space of bounded operators from E_2 to E_i'' ($E_o'' = e$, $E_1'' = E_1'$, $E_2'' = E_2$). We shall look for an R of the form

$$R(x_o,x,y) = r(x_o,x,y)y ,$$

where r is a continuous linear form on E_2. We localize the problem by multiplying r with a C^∞ function of compact support which is constant near the origin, which we call χ. Let $\chi(x,y) = 0$ if $\|x\| + \|y\| > \eta$, and $\chi(x,y) = 1$ if $\|x\| + \|y\| < \eta/2$.

If r satisfies the equation

$$\lambda r(x_o,x,y) = \chi(z)\ L_o(z) \tag{12.3}$$

$$+\ \chi(z)\ r(\lambda x_o + L_o(z)y,\ A_1'x + L_1(z)y,\ A_2y + L_2(z)y)\ (A_2 + L_2(z))\ ,$$

where $z = (x_o,x,y)$, then the corresponding R satisfies Eq.(12.2) local-ly. Let L be the linear space of functions f from E_2 to E_2^* such that

$$|||f||| = \text{Max}\Big(\ \sup_{||z-z'||<\eta}\ ||f(z)-f(z')||_{E_2^*}\ /\ ||z-z'||^\eta\ ,$$

$$\sup_{z\,\in\,E}\ ||f(z)||_{E_2^*}\ /\ ||z||\ \Big)\qquad.$$

This defines a norm on L , and with this norm, L is a Banach space. Since L_o is C^k with compact support, we have $\lambda^{-1}\chi L_o \in L$. Let τ be the linear operator given by

$$\tau f(z) = \lambda^{-1}\chi(z)\ f(\lambda x_o + L_o(z)\ y,A_1'x + L_1(z)y,A_2y + L_2(z)y)\ (A_2 + L_2(z)).$$

A tedious but straightforward substitution of the definitions shows that τ is a contraction on L, provided η is sufficiently small ($O(\varepsilon)$ is sufficient). The most marginal term in this estimate comes from the second norm in the definition of $|||.|||$ and is $\lambda^{-1}\ (\lambda+O(\eta))\lambda_4$, which is less than one provided η is small. Since $\lambda^{-1}\chi L_o \in L$, the equation

$$(I - \tau)\ r = \lambda^{-1}\chi L_o$$

has a unique solution, and this solves (12.3). In order to prove that r is C^1, let us remark that the following iteration scheme converges to r:

$$X_o = 0,$$

$$X_n = \tau\ X_{n-1}\ +\ \lambda^{-1}\chi L_o\ ,\quad n \geq 1\quad.$$

Since L_o is c^k, we have recursively that X_n is c^1, and moreover

$$DX_n(z) = \tau' \, DX_{n-1}(z) + \lambda^{-1} \, D\big(\chi(z) L_o(z)\big)$$

$$+ \lambda^{-1} \, X_{n-1}(t'z) \, \big(D\chi(z) \, (A_2 + L_2(z)) + \chi(z) \, DL_2(z)\big) \quad ,$$

where τ' is given by

$$(\tau' \, g)(x_o, x, y)$$

$$= \lambda^{-1} \, \chi(z) \, g(\lambda x_o + L_o(z)y, A_1'x + L_1(z)y, A_2y + L_2(z)y)$$

$$(A_2 + L_2(z) \, , \, Dt'(z)) \quad .$$

Here $g(z) \in (E_2 \times E)^*$, and $\tau'g(z) \in (E_2 \times E)^*$. We define the Banach space L' to be the linear space of functions g from E to $(E_2 \times E)^*$, equipped with the norm

$$|||g||| = \text{Max} \Big(\sup_{||z-z'|| < \mu} ||g(z) - g(z')||_{(E_2 \times E)^*} \, / \, ||z - z'||^\mu \, ,$$

$$\sup_{z \in E} ||g(z)||_{(E_2 \times E)^*} \, / \, ||z||^\mu \Big) \quad .$$

One checks that $\lambda^{-1} \, D(\chi L_o)$ is in L' since this function is c^k with compact support.

A direct verification, using the definitions shows that for $\mu > 0$ sufficiently small (e.g. $O(\varepsilon^2)$ is sufficient), the map τ' is a contraction on L'. If we define $Y_n = DX_n$ and

$$B_n = \lambda^{-1} \, X_n \circ t' \, (\chi \, DL_2 + (A_2 + L_2)D\chi) + \lambda^{-1} \, D(\chi L_o) \quad ,$$

then we have

$$Y_n = \sum_{j=0}^{n-1} (\tau')^j \; B_{n-j} \qquad .$$

Since $X_n \to r$ in L, we find

$$B_n(z) \to B_\infty(z) = \lambda^{-1} \; r(t(z))(\chi(z) \; DL_2(z) + D\chi(z)(A_2 + L_2(z)))$$

$$+ \lambda^{-1} \; D(\chi \; L_o)(z)$$

in L'. We may write

$$Y_n = \sum_{j=0}^{n-1} (\tau')^j \; B_\infty + \sum_{j=0}^{n-1} (\tau')^j \; (B_{n-j} - B_\infty) \; ,$$

and since τ is a contraction, the last term goes to zero like a power and we find

$$Y_n \to \sum_{j=0}^{\infty} (\tau')^j \; B_\infty \qquad \text{in } L'.$$

Since the topology of L' implies uniform convergence, we find that r is C^1 and $Dr = \lim_{n \to \infty} DX_n$ in L'. The existence and differentiability of U^{-1} follows.

Note concerning Theorem 4.4 and Corollary 4.5:

The statements as made in Part I are only conjectures in the form in which they have been written. In this section we have only proved that there is a C^1 transformation which diagonalizes the most unstable direction alone. This is sufficient for our purpose, except for the calculation of the critical index in Eq. (5.35), where we need a diagonalization in two variables. This could be done in a similar fashion as above, again with a C^1 transformation by doing perturbation theory for the eigenvalues up to order $O(\varepsilon^2)$ first, and handling the subspace for λ_4 on a separate basis.

13. Crossover. Part I

In this section, we control the flow in a neighborhood of the critical fixed point for $\beta \neq \beta_{crit}$. In this region, we shall use the normal form of the flow which we derived in Section 12. The single spin distribution is supposed to be of the form $\phi_\varepsilon + f$ in the discussion below. We shall work throughout with the unnormalized spin distributions. The transition to the normalized case will be easy.

The normal form of the flow obtained in Section 12 applies in a region whose size is conveniently described by the norms $|\cdot|_j$, j=1,2 of Mather. By (12.1), we have

$$\|f\|_\infty \leq |f|_j \leq \|f\|_\infty \varepsilon^{-14} \quad . \qquad\qquad (13.1)$$

We may assume that the various restrictions on the size of the neighborhood in which we did our calculations lead to the conclusion that the normal form holds in a ball of radius ε^{90} in L_∞ , provided $\varepsilon > 0$ is sufficiently small. (It is certainly obvious from what we have done so far that this neighborhood does shrink at most as a power of ε, and it is only this fact which is of importance in the sequel.)

We write

$$T(f) \;=\; \mathcal{N}_\varepsilon^0 (\phi_\varepsilon + f) \;-\; \phi_\varepsilon \;.$$

If $\|f\|_\infty \leq \varepsilon^{90}$, then the normal form of $T^n f$ holds as long as $\|T^n f\|_\infty \leq \varepsilon^{90}$. We consider now $f = a_o e_o + a_2 e_2 + r$, where e_o and e_2 are the first two eigenvectors of $\mathcal{A}_{\phi_\varepsilon}$ on the even subspace of L_∞, $e_o = \phi_\varepsilon$, and r is a (small) remainder. Neglecting for the moment the action of S from Theorem 4.2, we have something like

$$T^n(f) \;\sim\; 2^n\, a_o e_o \;+\; \lambda_2^n\, a_2 e_2 \;+\; r_n \;. \tag{13.3}$$

Thus the vector $\phi_\varepsilon + f$ is mapped to $\phi_\varepsilon (1+2^n a_o) + \lambda_2^n\, a_2 e_2 + r_n$, which by a scalar change can be brought to the form

$$\phi_\varepsilon \;+\; \frac{\lambda_2^n}{1+2^n a_o}\, a_2 e_2 \;+\; \frac{1}{1+2^n a_o}\, r_n \;. \tag{13.4}$$

But scalar changes do not show up when we calculate normalized expectation values, and so (13.4) is just as useful as (13.3). We now formulate the statement (13.3) precisely. Let $\phi(\beta,.)$ be a temperature trajectory as in Lemma 5.2. Then for β near β_{crit}, we have $\phi(\beta,.)=\phi_\varepsilon+f' \notin \hat{\mathcal{W}}_s$ and

$$\hat{S}^{-1} f' \;=\; a_2' \hat{e}_2 \oplus r' \;, \qquad \begin{cases} a_2' > 0 \text{ if } \beta < \beta_{crit} \\ a_2' < 0 \text{ if } \beta > \beta_{crit} \end{cases} \;.$$

From the properties derived for $\hat{U}$ in Section 12, we have with the same signs

$$f \;=\; \hat{U}^{-1} \hat{S}^{-1} f' \;=\; a_2 \hat{e}_2 \oplus r \;. \tag{13.5}$$

THEOREM 13.1. Let f be as in (13.5), and let $\|f\|_\infty < \varepsilon^{330}$. Then for some $n < \infty$ the iterate

$$f_n \;=\; \hat{T}^n (f)$$

satisfies

$$S f_n = a_2^{(n)} \hat{e}_2 \oplus r_n , \qquad (13.6)$$

with
$$\varepsilon^{100} < |a_2^{(n)}| < 3 \varepsilon^{100} ,$$

and
$$\|r_n\|_\infty \leq \varepsilon^{308} .$$

COROLLARY 13.2. For some $n < \infty$ one has

$$\hat{\mathcal{N}}_\varepsilon (\phi_\varepsilon + f') = \text{const.} (\phi_\varepsilon + a_2^{(n)} \hat{e}_2 + r_n) ,$$

with the same bounds as in Theorem 13.1, but with $\|r_n\|_\infty \leqslant \varepsilon^{137}$.

Proof: This follows because SU is near 1, and in fact $SU - 1$ is $3/2$-
Lipshitz of norm ε^2.

Proof of Theorem 13.1: This follows by the construction of Section
12 applied to $\hat{\mathcal{N}}$ on the even subspace (there is only one unstable direction!).

Note: If $\phi_\varepsilon + f \notin \mathcal{W}_s$ is positive, then $\mathcal{N}_\varepsilon^n (\phi_\varepsilon + f)$ has the same form

$$\text{const.} (\phi_\varepsilon + a_2^{(n)} e_2 + r_n) ,$$

with bounds as above, since the normalized and the unnormalized transformations differ only by a finite, non-zero factor.

We now push these estimates further. We shall need in the follow-
ing some more detailed information about the eigenvectors e_{2j} of $\mathcal{A}_{\phi_\varepsilon}$
in L_∞ . We start by studying this in perturbation theory for a function

f of Eq. (8.1). We fix N and we let f_ε be such that $\mathcal{N}(f_\varepsilon) - f_\varepsilon$

$= O(\varepsilon^{26\,N\,+\,40}\exp(-\varepsilon\theta x^4/2))$. Then we can approximate the eigenvectors

of $\mathcal{A}_{f_\varepsilon}$ in the following sense.

LEMMA 13.3. Fix $L > 0$ and $K > 0$. Then there is for all $p \leq K$ a polynomial

$$h_{2p,K}(x) = H_{2p}(\gamma^{\frac12}x) + \sum_{j=1}^{K}\;\sum_{\ell \leq j+p} a_{j\ell}^{(p)}\;\varepsilon^j\;x^{2\ell}, \qquad (13.7)$$

and a polynomial $\lambda'_{2p}(\varepsilon)$, such that if $f_{\varepsilon,L}(x) = \exp(-\varepsilon\theta x^4)\;P_L(\varepsilon,x)$,

(cf. (3.19)), then

$$| \mathcal{A}_{f_{\varepsilon,L}}(h_{2p,K}(x)\exp(-\varepsilon\theta x^4))(z) - \lambda'_{2p}\,h_{2p,K}(z)\exp(-\varepsilon\theta z^4)|$$

$$\leq \quad \varepsilon^{(K-p)/2}\;\exp(-\varepsilon\theta z^4/2) \quad . \qquad (13.8)$$

Proof: If we group the exponential factors together, the equation for

the eigenvalue λ''_{2p} and the eigenvector $v_{2p}(x)\exp(-\varepsilon\theta x^4)$ of $\mathcal{A}_{f_{\varepsilon,L}}$ is

$$\int \exp(-(2/c^2-1)\,\varepsilon\theta x^4)\exp(-u^2(1+12\varepsilon\theta x^2/c))\,P_L(\varepsilon,xc^{-\frac12}-u)\,v_{2p}(xc^{-\frac12}+u)\,du$$
$$\cdot\exp(-2\varepsilon\theta u^4)$$

$$= \quad \lambda''_{2p}\;v_{2p}(x) \quad . \qquad (13.9)$$

These equations can be solved recursively, as described in the program

to compute λ_2 (Section 9), by setting as initial values $\lambda''_{2p} = 2\,c^{-p}$,

$v_{2p}(x) = H_{2p}(\gamma^{\frac12}x)$, and observing that the operator on the LHS of

(13.9) is $\mathcal{A}_1$ + higher orders in ε. By our analysis of Section 8, as a

formal power series, this operator contains only terms of the form

$\varepsilon^j\,x^k$ with $k \leq 2j$, hence the form (13.7) of any perturbative solution.

The bound (13.8) follows now from the fact that the LHS is bounded by

$\exp(-\varepsilon\theta z^4/2)$ times

$$C_{K,L}\;\sum_{j=K+1}^{K+L} \varepsilon^j\;(1 + z^{2j+2p})\;\exp(-\varepsilon\theta z^4/2) \quad . \qquad (13.10)$$

We note now a useful inequality which we call E-estimate (exponential

estimate): If $\theta > 0$, then

$$|x^{2k} \exp(-\epsilon\theta x^4/2)| \leq O(\epsilon^{-k/2}) , \quad k = 0,1,2,\ldots \quad .$$

Therefore the term (13.10) is bounded by

$$C_{K,L} \sum_{j=K+1}^{K+L} \epsilon^j \epsilon^{-j/2-p/2} \leq \epsilon^{(K-p)/2} ,$$

uniformly for $K \leq K_o$, $L \leq L_o$, for sufficiently small ϵ. This comple-

tes the proof of the lemma.

In the sequel we assume $L \geq K$. Then the function $h_{2p}(x)$ is corr-

ect in perturbation theory up to order K in ϵ. We now want to show that

$h_{2p,K}(x) \exp(-\epsilon\theta x^4)$ is near to the true eigenfunction $v_{2p}(x) \cdot$

$\exp(-\epsilon\theta x^4)$ and that λ'_{2p} is near to the true eigenvalue λ''_{2p} of $\mathcal{A}_{f_{\epsilon,L}}$.

LEMMA 13.4. For each $p \leq (K-10)/5$ there is an eigenvector $v_{2p}(x) \cdot$

$\exp(-\epsilon\theta x^4)$ of $\mathcal{A}_{f_{\epsilon,L}}$ such that

$$|(v_{2p}(x) - h_{2p,K}(x)) \exp(-\epsilon\theta x^4)| \leq \epsilon^{(K-5p)/2-5} \exp(-\epsilon\theta x^4/2) ,$$

and (13.11)

$$|\lambda'_{2p} - \lambda''_{2p}| \leq \epsilon^{(K-3p)/2-3} .$$

Proof: Since the ideas are analogous to those used in the proof of Theo-

rem 10.2, we only give some hints. One solves the eigenvalue equation

for a remainder term $s_{2p} \exp(\epsilon\theta x^4/2)$, where s_{2p} is the difference of

the LHS of Equation (13.11). It is

$$s_{2p} \exp(\epsilon\theta x^4/2)$$

(13.12)

$$= (\mathcal{B}_{f_{\epsilon,L}} - \lambda'_{2p})^{-1} \{(\lambda''_{2p} - \lambda'_{2p}) h_{2p} \exp(-\epsilon\theta x^4/2) + O(\epsilon^{(K-p)/2})\} .$$

We shall choose $s_{2p} \exp(\varepsilon\theta x^4/2)$ in $V_p^\perp \subset L_\infty$, the spectral subspace

of $\mathcal{B}_{f_{\varepsilon,L}}$ corresponding to the complement of λ_{2p}''. As in the proof of

Theorem 10.1, we find that the corresponding projection has norm boun-

ded by $O((4c^p/2)^{n_0}) = O((c^{p+2})^{n_0}) = O(\varepsilon^{-p-2})$, and hence the ex-

pression $\{\ldots\}$ is in $V_p^\perp$ for some λ_{2p}'' with $|\lambda_{2p}'' - \lambda_{2p}'| \le$

$O(\varepsilon^{(K-p)/2-p-2})$. On the other hand,

$$\| ((\mathcal{B}_{f_{\varepsilon,L}} - \lambda_p') |_{V_p^\perp})^{-1} \| \le O(\varepsilon^{-p-2}) \quad,$$

so that the result follows. Finally, we go back to the operator of int-

erest, namely $\mathcal{A}_{\phi_\varepsilon}$.

COROLLARY 13.5. The eigenvectors e_{2p} and the eigenvalues λ_{2p} of $\mathcal{A}_{\phi_\varepsilon}$

satisfy for $p \le (K-10)/5$,

$$|e_{2p}(x) - h_{2p}(x) \exp(-\varepsilon\theta x^4)| \le \varepsilon^{(K-5p)/2-5} \exp(-\varepsilon\theta x^4/2) \quad,$$

and

$$|\lambda_{2p} - \lambda_{2p}'| \le \varepsilon^{(K-3p)/2-3} \quad.$$

Proof: This follows by standard perturbation theory. Choose L in Lemma

13.3 so large that $L \ge K$ (see below) and such that $\|\phi_\varepsilon - f_{\varepsilon,L}\|_\infty \le O(\varepsilon^K)$.

Then the assertion follows from Lemma 13.4.

COROLLARY 13.6. For $p,q \le M$ one has in L_∞ the identity

$$\mathcal{N}(e_{2p}, e_{2q}) = \sum_{\ell=0}^{M} \dot{c}_{pq\ell}(\varepsilon) \, e_{2\ell} + O(\varepsilon^{M/2-p-q}). \tag{13.13}$$

If $\ell > p+q$, then one has $c_{pq\ell}(\varepsilon) = O(\varepsilon^{\ell-p-q})$. $\tag{13.14}$

Proof: Let K be so large that $(K - 5M)/2 - 5 \ge (M + 1)/2$. $\tag{13.15}$

By Corollary 13.5, we may write

$$e_{2p}(x) = Q_{2M}^{(p)}(\varepsilon,x) \exp(-\varepsilon\theta x^4) + \sum_{j=M+1-p}^{K} \varepsilon^j \sum_{n=0}^{j+p} d_{njp} x^{2n} \exp(-\varepsilon\theta x^4)$$

$$+ \quad O(\varepsilon^{(K-5p)/2-5} \exp(-\varepsilon\theta x^4/2)) \ ,$$

where $Q_{2M}^{(p)}$ is that part of h_{2p} which contains only terms of degree up to $2M$ in x. By the E-estimate, and by (13.15),

$$e_{2p}(x) = Q_{2M}^{(p)}(\varepsilon,x) \exp(-\varepsilon\theta x^4) + O(\varepsilon^{(M+1)/2-p}). \tag{13.16}$$

We now apply $\mathcal{N}$ to e_{2p} written in the form (13.16). Then

$$\mathcal{N}(e_{2p},e_{2q})(x) = \sum_{\ell=0}^{M} c_{pq\ell}(\varepsilon) Q_{2M}^{(\ell)}(\varepsilon,x) \exp(-\varepsilon\theta x^4) + r \ , \tag{13.17}$$

and r collects the various remainder terms. By construction, the poly-nomials which are not absorbed in the sum in (13.17) have terms of the form $\varepsilon^{m-p-q} x^{2m'}$ with $m \geq m' > M$. Together with $\exp(-\varepsilon\theta x^4)$ this yields a bound $O(\varepsilon^{(M+1)/2-p-q})$. Similar considerations apply for the other terms (crossed terms between the first and the second term in (13.16)) and yield a bound $O(\varepsilon^{(M+1)/2-p-q})$. Finally,

$$Q_{2M}^{(\ell)}(\varepsilon,x) \exp(-\varepsilon\theta x^4) = e_{2\ell} + O(\varepsilon^{(M+1)/2-\ell}) \ ,$$

so that (13.13) is proved, since $c_{pq\ell} = O(\varepsilon^{\ell-p-q})$, by perturbation theory.

We go back to controlling the flow. By Corollary 13.2 we have to consider a function of the form

$$f = b_2' e_2 + r' \ ,$$

with $\varepsilon^{100} < |b_2'|$, $\|r'\|_\infty \leq \varepsilon^{137}$ By Corollary 13.5, f can be represented in the form

$$f = b_2 e_2 + \sum_{k=2}^{K'} b_{2k} e_{2k} + s \quad , \quad K' = 5 \quad ,$$

where s is in the spectral subspace V corresponding to the eigenvalues λ_{2K+2} , λ_{2K+4} , ..., and one has $|b_2 - b_2'| < \varepsilon^{137}$, $|b_j| < \varepsilon^{137}$, if $j \neq 2$ and finally $\|s\|_\infty < \varepsilon^{134}$.

We shall bound recursively $T^n(f)$, which we write in the form

$$T^n(f) = \sum_{j=1}^{K'} b_{2j}^{(n)} e_{2j} + \mathcal{A}^n(s) + r^{(n)} \quad .$$

We shall do a total of n_2 steps where n_2 is such that $|b_2^{(n_2)}| < \varepsilon^{15/16}$, i.e. $\lambda_2^{n_2} |b_2^{(0)}| \sim \varepsilon^{15/16}$. Our aim is to show that the other terms remain an order of magnitude smaller. We do estimates recursively. Recall that $n_0 = \log(2\gamma/\varepsilon + 1)$. First we observe that by Theorem 10.1, one has

$$\| \mathcal{A}^n(s) \|_\infty \leq \begin{cases} \varepsilon^{127} & , \text{ if } n \leq n_0 \ , \\ \\ \varepsilon^{127} (2/c^k)^{n-n_0}, & \text{ if } n > n_0 \ . \end{cases} \tag{13.18}$$

The inductive bounds we are proving are as follows. Let β be slightly smaller than 15/16 and let β' be slightly smaller than β . Let λ_{2j+} denote the number $\lambda_{2j} + \varepsilon^{\frac{1}{4}}$. Then

$$|b_2^{(n)}| \leq \lambda_{2+}^n \varepsilon^{100} \quad , \tag{13.19}$$

$$|b_{2j}^{(n)}| \leq \max\left(\lambda_{2j+}^n \varepsilon^{137}, \varepsilon^{\beta' j - 2\beta} (\lambda_{2+}^{n-1} \varepsilon^{100})^2 \right) \quad , \tag{13.20}$$
$$j \neq 1,$$

$$\|r^{(n)}\|_\infty \leq (2 + \varepsilon^{\frac{1}{4}})^n \varepsilon^{202} \quad . \tag{13.21}$$

These inequalities are obviously true for n=0, by our previous considerations. To perform an induction step, we compute

$$T^{n+1}(f) = \sum_{j=1}^{K} b_{2j}^{(n)} \lambda_{2j} e_{2j} + \sum_{j,k} b_{2j}^{(n)} b_{2k}^{(n)} \mathcal{N}(e_{2j}, e_{2k})$$

$$+ \mathcal{A}^{n+1}(s) + \mathcal{A}(r^{(n)}) + 2 \mathcal{N}(\sum_{j=1}^{K} b_{2j}^{(n)} e_{2j}, \mathcal{A}^n(s))$$

$$+ 2 \mathcal{N}(\sum_{j=1}^{K} b_{2j}^{(n)} e_{2j}, r^{(n)}) + 2 \mathcal{N}(\mathcal{A}^n(s), r^{(n)})$$

$$+ \mathcal{N}(\mathcal{A}^n(s)) + \mathcal{N}(r^{(n)})$$

$$= X_1 + \ldots + X_9 .$$

The contributions to $r^{(n+1)}$ come from $X_2 + X_4 + X_5 + X_6 + X_7 + X_8 + X_9$, and are easily seen to fulfill the recursive condition (13.20). The Equations (13.18), (13.19) follow from Corollary 13.6, and by analyzing the identity

$$b_{2j}^{(n+1)} = \lambda_{2j} b_{2j}^{(n)} + \sum_{k,\ell} c_{k\ell j} b_{2k}^{(n)} b_{2\ell}^{(n)} , \qquad (13.22)$$

using that $c_{k\ell j} = O(\varepsilon^{j-k-\ell})$ if $j \geq k+\ell$. The "maximum" in (13.19) takes into account whether the major contribution to $b_{2j}^{(n+1)}$ comes from the first term or from the second term on the RHS of Eq.(13.22). The rest is simple arithmetics. By rearranging the terms we get the following result.

THEOREM 13.7. For some $n_2 < \infty$ the iterate $\mathcal{N}^{n_2}(\phi_\varepsilon + f)$ is of the form

$$\mathcal{N}^{n_2}(\phi_\varepsilon + f)$$

$$= \text{const.} \left(\phi_\varepsilon \pm d\varepsilon^{15/16} e_2 + \sum_{j=2}^{5} O(\varepsilon^{29j/32}) e_{2j} \right.$$

$$\left. + O(\varepsilon^{19/8}) \exp(-\varepsilon\theta x^4/2) \right) , \qquad (13.23)$$

where $1 \leq d \leq 2$.

We have included the exponential factor without having carried it explicitly through the preceding calculations.

14. Crossover, Part II. A) Single Phase Region

We now treat the flow on a different footing, depending on the sign of the e_2-term in Theorem 13.7. In this section, we take the negative sign (i.e. the coefficient of the x^2-term is negative). We first rewrite the result of the action $T^{n_2}(f)$ in exponential form.

THEOREM 14.1. <u>There is a constant</u> $\alpha = \theta + O(\varepsilon^{7/8})$ <u>and a constant</u> $b > 0$ (<u>not depending on</u> ε) <u>such that</u>

$$\mathcal{N}^{n_2}(\phi_\varepsilon + f)(x) = \text{const.} \Big(\exp(-\alpha\varepsilon x^4 - b\varepsilon^{15/16} x^2)$$

$$\cdot(1 + \sum_{j=3}^{5} k_j(\varepsilon) x^{2j} \varepsilon^{29j/32}) + \exp(-\alpha\varepsilon x^4/2) R_2'(x) \Big) ,$$

(14.1)

<u>where</u> $|R_2'(x)| \leq O(\varepsilon^{19/8})$.

<u>Proof</u>: By Eq.(13.7) and Corollary 13.5 we may expand each of the e_{2j} as a polynomial plus a remainder. For example,

$$e_6(x) = \big(H_6(\gamma^{\frac{1}{2}}x) + \varepsilon \cdot \text{polynomial of degree 8 in } x + \varepsilon^2 \cdot \dots \big)$$

$$\cdot \exp(-\varepsilon\theta x^4) + O(\varepsilon^{-\frac{1}{2}}) \exp(-\varepsilon\theta x^4/2) .$$

Using also the ε-expansion for ϕ_ε, we can write the result of (13.23) in exponential form, using e.g.

$$\exp(-\varepsilon x^4) (1 - ax^2)$$

$$= \exp(-\varepsilon x^4) \exp(-ax^2) (1 - a^2 x^4/2! - a^3 x^6 2/3! - a^4 x^8 3/4!)$$

$$+ \exp(-\varepsilon x^4/2) O(a^5 \varepsilon^{-5/2}) .$$

By the E-estimate (page 140) the assertion of the theorem follows from Theorem 13.7.

We now perform some more iteration steps until the quadratic part in (14.1) has a coefficient of about $\varepsilon^{\frac{1}{2}}$ (after the first n_2 steps it was about $\varepsilon^{15/16}$). The control of these steps is provided by the

LEMMA 14.2. If g is of the form

$$g(x) = \exp(-\varepsilon Ax^4 - Bx^2) (1 + \sum_{j=3}^{5} c_j \varepsilon^{29j/32} x^{2j})$$

$$+ \exp(-\varepsilon ax^4/2) R(x) ,$$

with $\varepsilon^{15/16} < B < \varepsilon^{\frac{1}{2}}$, $|c_j| < C_o$, $|R| < \varepsilon^{11/8}$, then $\mathcal{N}(g)$ has the same representation up to a constant factor (primed quantities), with $A' = (2/c^2)A + O(\varepsilon)$, $B' = (2c^{-1} + O(\varepsilon))B$, $|R'| \leq 2(1 + O(\varepsilon^{1/16}))|R|$ uniformly in C_o .

Proof: Since the first term in g is bounded by $(1+\varepsilon)$, it suffices to look at the action of $\mathcal{N}$ on this first term g_1 (while the crossed term and the remaining terms will yield a first contribution to R'). To control the action of $\mathcal{N}$, we write its definition:

$$\mathcal{N}(g_1)(z) = \exp(-2c^{-2}\varepsilon Az^4 - 2c^{-1}Bz^2)$$

$$\cdot \pi^{-\frac{1}{2}} \int du \, \exp(-u^2(1+2B)) \exp(-12u^2 A\varepsilon z^2/c) \exp(-2\varepsilon\theta u^4)$$

$$\cdot \left(1 + \sum_{j=3}^{5} c_j \varepsilon^{29j/32} (zc^{-\frac{1}{2}} - u)^{2j}\right)$$

$$\cdot \left(1 + \sum_{j=3}^{5} c_j \varepsilon^{29j/32} (zc^{-\frac{1}{2}} + u)^{2j}\right) ,$$

which, upon expansion of the last exponential and the powers 6, 8, 10, becomes

$$\exp(2c^{-2}\varepsilon Az^4 - 2c^{-1}Bz^2)\ (1+2B)^{-\frac{1}{2}}$$

$$\cdot\left(1 + O(\varepsilon) + O(\varepsilon z^2) + \sum_{j=3}^{5}\ (2c^{-j}\ c_j\ \varepsilon^{29j/32}\ z^{2j} + O(\varepsilon^j z^{2j})\ )\right)$$

+ negligeable terms. $\hspace{4cm}$ (14.2)

The result follows at once, noting that there is a contribution $O(\varepsilon^{1/16})$ to R', and absorbing the low order terms into the exponential, see also the proof of Proposition 15.4.

COROLLARY 14.3. For some n_3 , one has

$$\mathcal{N}^{n_3}(\ \phi_\varepsilon + f\ )(x)$$

$$= \text{const.}\ \left(\exp(-\rho\varepsilon^{\frac{1}{2}}x^2 - a''\varepsilon x^4) + O(\varepsilon^{11/8})\ \exp(-\rho\varepsilon^{\frac{1}{2}}x^2)\right)\ ,\hspace{1cm}(14.3)$$

with $1 \le \rho \le 2$, and with $a'' = O(1)$.

Proof: By the control of the coefficient of x^2, we find $(2/c)^{n_3-n_2} \simeq \varepsilon^{\frac{1}{2}-15/16}$. Therefore, by Theorem 14.1 and Lemma 14.2, the remainder term is bounded by

$$2^{n_3-n_2}\ O(\varepsilon^{19/8})\ \le\ (\varepsilon^{-7/16})^{2+O(\varepsilon)}\ O(\varepsilon^{19/8})\hspace{2cm}(14.4)$$

$$\le\ O(\varepsilon^{12/8})\ \ ,$$

since $2/c = 2^{\frac{1}{2}} + O(\varepsilon)$. The assertion follows now by absorbing the polynomial terms into the remainder, using the E-estimate. Note that we have got rid of all polynomial terms in $\mathcal{N}^{n_3}(\phi_\varepsilon + f)$. We now do some more steps to accentuate the Gaussian part.

PROPOSITION 14.4. For some n_4, one has

$$\mathcal{N}^{n_4}(\phi_\varepsilon + f)(x) = \text{const.}\left(\exp(-\rho\varepsilon^{1/6} x^2)\, (1 + R'(x)) \right) , \qquad (14.5)$$

with $1 \le \rho \le 2$, and $|R'| < O(\varepsilon^{2/3})$.

Proof: We first use the following fact:

If $\mathcal{N}^n(\phi_\varepsilon + f)$ is of the form (14.3), i.e.

$$\mathcal{N}^n(\phi_\varepsilon + f)(x) = \exp(-A\varepsilon^{\frac{1}{2}}x^2 - B\varepsilon x^4) + R(x)\, \exp(-A\varepsilon^{\frac{1}{2}}x^2) ,$$

with $|R| \le O(\varepsilon^{11/8})$, then $\mathcal{N}^{n+1}(\phi_\varepsilon + f)$ is of the same form (primed quantities), with $B' = 2B/c^2 + O(\varepsilon)$, $A' = 2A/c + O(\varepsilon)$, and $|R'| \le (2 + O(\varepsilon^{1/16}))|R|$.

This is seen as in the proof of Lemma 14.2, with the difference that the polynomial terms are absorbed into the exponential (up to order 4) and into the remainder (higher order terms, starting from order 6). We get thus the result (14.5) by a calculation similar to (14.4), after expanding $\exp(-B\varepsilon x^4)$.

Now, nothing is in our way to let (by further iterations) the image of $\phi_\varepsilon + f$ tend to a δ-function. Our next intermediary result is

COROLLARY 14.5. For some n_5 one has

$$\mathcal{N}^{n_5}(\phi_\varepsilon + f)(x) = \text{const.}\left(\exp(-Ax^2)\, (1 + R_5(x)) \right) ,$$

with $10^4 < A < 10^5$, $|R_5| < \varepsilon^{\frac{1}{4}}$.

Proof: As before, the bound on R_5 follows from (14.2).

From this point onward, it is convenient to analyze the iteration scheme in a slightly different way. Let $g_0(x) = \mathcal{N}^{n_5}(\phi_\varepsilon + f)(x)$, and define

$$g_n(x) = \mathcal{N}^n(g_0)((c/2)^{n/2} x) \quad . \tag{14.6}$$

A comparison with the scaling relation (5.25) for the susceptibility χ, $\chi_{2N,g} = (2/c) \chi_{N,\mathcal{N}(g)}$ suggests that the functions g_N converge (as the density of a measure) to a fixed function g_∞ . In fact we show now that g_∞ is a Gaussian. We proceed by an inductive construction, as before, and it is easy to check that the inductive assumption given below holds for the function const. $\cdot g_0$. Note also that the definition (14.6) implies

$$g_{n+1}(x) = \pi^{-\frac{1}{2}} (c/2)^{n/2} \int du \, \exp(-u^2(c/2)^n) \, g_n(2^{-\frac{1}{2}}x-u) \, g_n(2^{-\frac{1}{2}}x+u). \tag{14.7}$$

THEOREM 14.6. Suppose that

$$g_n(x) = (c/2)^{2^{n-1}-n/2-\frac{1}{2}} \exp(-\alpha_n x^2/2) (1 + R_n(\alpha_n^{\frac{1}{2}} x)) \quad ,$$

with $\alpha_0 > 10^4$, and suppose that for

$$\gamma_n = 1/2 - K'(c/2)^{2n} \quad , \quad 3K' < \alpha_0^{-1} \quad ,$$

one has

 (i) R_n is orthogonal to 1 and x^2 in L_{2,γ_n} ,
 (ii) $\|R_n\|_{2,\gamma_n} \leq K(c/2)^n$, $K = 10^{-4}$.

Then g_{n+1} has the analogous properties with n replaced by n+1 and in addition $|\alpha_{n+1} - \alpha_n| = O((c/2)^{2n})$.

Proof: Using (14.7), we may write

$$g_{n+1}(x) = (c/2)^{2^n - (n+1)/2 - \frac{1}{2}} \exp(-\alpha_n x^2/2) \, \alpha_n^{-\frac{1}{2}} \, h_{n+1}(\alpha_n^{\frac{1}{2}} x) \quad ,$$

where

$$h_{n+1}(x) = \pi^{-\frac{1}{2}} \int du \; e^{-u^2(1+(c/2)^n/\alpha_n)} \, (1+R_n(2^{-\frac{1}{2}}x+u))(1+R_n(2^{-\frac{1}{2}}x-u)).$$

By the orthogonality condition on R_n we have for $R_n'(x) = R_n((2\gamma_n)^{-\frac{1}{2}}x)$
the relations $\|R_n'\|_{2,\frac{1}{2}} = \|R_n\|_{2,\gamma_n}$, and R_n' is orthogonal to 1
and x^2 in $L_{2,\frac{1}{2}}$. Therefore we shall write

$$(2\gamma_n)^{-\frac{1}{2}} \, h_{n+1}((2\gamma_n)^{-\frac{1}{2}} x)$$

$$= \pi^{-\frac{1}{2}} \int du \; e^{-2\gamma_n u^2(1+(c/2)^n/\alpha_n)} \, (1+R_n'(2^{-\frac{1}{2}}x+u))(1+R_n'(2^{-\frac{1}{2}}x-u))$$

$$= \pi^{-\frac{1}{2}} \left[\int du \, \{\exp(-2\gamma_n u^2 (1+(c/2)^n/\alpha_n)) - \exp(-u^2)\} \right.$$

$$\left. \cdot R_n'(2^{-\frac{1}{2}}x - u) \, (2 + R_n'(2^{-\frac{1}{2}}x + u)) \right]$$

$$+ (2\gamma_n (1 + (c/2)^n/\alpha_n))^{-\frac{1}{2}} \left(1 + \mathcal{A}_{c=2}(R_n')(x) + \mathcal{N}_{c=2}(R_n')(x)\right)$$

$$= (2\gamma_n (1 + (c/2)^n/\alpha_n))^{-\frac{1}{2}} \left(1 + \mathcal{A}_{c=2}(R_n')(x) + r_{n+1}(x)\right) \quad .$$

The main point of these manipulations is that $\mathcal{A}_{c=2}$ (which is our usual
linearized operator for c=2) has the property that $\mathcal{A}_{c=2}(R_n')(x)$ is
orthogonal to 1 and to x^2 in $L_{2,\frac{1}{2}}$ (this is easily checked from the definitions). Hence

$$h_{n+1}(t) = (1 + (c/2)^n/\alpha_n)^{-\frac{1}{2}} (1 + \mathcal{A}_2(R_n')((2\gamma_n)^{\frac{1}{2}}t) + r_{n+1}((2\gamma_n)^{\frac{1}{2}}t)),$$

and by the orthogonality, we have

$$\|\mathcal{A}_2(R_n')((2\gamma_n)^{\frac{1}{2}}\cdot)\|_{2,\gamma_n} = \|\mathcal{A}_2(R_n')\|_{2,\frac{1}{2}}$$

$$\tag{14.8}$$

$$\leq \quad \tfrac{1}{2} \|R_n'\|_{2,\frac{1}{2}} = \tfrac{1}{2} \|R_n\|_{2,\gamma_n} \; .$$

Next, we bound $\|r_{n+1}\|_{2,\gamma_n}$. First of all, we note that

$$\| \mathscr{N}_{c=2}(f) \|_{2,\frac{1}{2}} \quad \leq \quad \text{const.} \, \| f \|_{2,\frac{1}{2}}^2 \, . \tag{14.9}$$

(This is a limiting case of Lemma 10.3 and it is proved as follows: If $f \in L_{2,\frac{1}{2}}$, we write $f(x) = \exp(x^2/4) \, f'(x)$, with $f' \in L_2(dx)$. Then

$$\| \mathscr{N}_2(f) \|_{2,\frac{1}{2}}^2 \quad = \quad \text{const.} \int dx \int du\,dv \, \exp(-u^2/2 - v^2/2)$$

$$\cdot |f'(x+u) \, f'(x-u) \, f'(x+v) \, f'(x-v)| \quad .$$

Now integrate over $v, x,$ and u .)

As in the case of $\mathcal{A}_2$, we deduce $\| \mathscr{N}_2(R'_n)((2\gamma_n)^{\frac{1}{2}} \cdot) \|_{2,\gamma_n} \leq$ const. $(2\gamma_n)^{\frac{1}{2}} \|R_n\|_{2,\gamma_n}^2$. It remains to bound the term with the difference of exponentials. The term linear in R'_n is bounded by use of the formula

$$\left| \int du \, \exp(-u^2) \, f(u) \, R'_n(2^{-\frac{1}{2}}z-u) \right|$$

$$\leq \left(\int du \, \exp(-u^2) \, |f(u)|^2 \int dv \, \exp(-v^2) \, |R'_n(2^{-\frac{1}{2}}z-v) \, R'_n(2^{-\frac{1}{2}}z+v)| \right)^{\frac{1}{2}},$$

and the Equations (14.9),(14.8). This yields a bound of about $(c/2)^{2n} \cdot$ const. $(K' + \alpha_n^{-1} + K'\alpha_n^{-1})$. The quadratic term is bounded using (14.9), which is true for any $L_{2,\lambda}$ norm.

We now set

$$h_{n+1}(x) \quad = \quad (1 + O(K(c/2)^{2n}) \,) \, \exp(-\delta_n x^2/2) \, (1+R_{n+1}((1+\delta_n)^{\frac{1}{2}}x) \,) \, ,$$

with $\delta_n = \alpha_{n+1}/\alpha_n - 1$ to be fixed such that R_{n+1} is orthogonal to 1 and to x^2 in $L_{2,\gamma_{n+1}}$. This orthogonality condition (for the second

Hermite polynomial) is after some manipulations seen to be equivalent
to the condition

$$\int dx \ ((1+\delta_n)(\gamma_{n+1}/\gamma_n)x^2 - 1) \ \exp(-(1+\delta_n)(\gamma_{n+1}/\gamma_n)x^2)$$

$$\cdot \exp(\delta_n x^2/(4\gamma_n)) \ (1 + \mathcal{A}_2(R_n')(x) + r_{n+1}(x)) \quad = \quad 0 . \quad (14.10)$$

Consider the LHS of (14.10) as a function $f(\delta_n)$ of δ_n. We want to solve
$f(\delta_n) = 0$, and we do this by the implicit function theorem. One gets

$$|f(0)| \quad \leq \quad \text{const.} \ (K + K') \ (c/2)^{2n} \ ,$$

$$\partial_\delta f(0) \quad = \quad 1 \ + \ O(K \ (c/2)^n) \ ,$$

$$|\partial_\delta^2 f(\delta)| \quad = \quad O(1) \ ,$$

as long as $|\delta| < K(c/2)^n$. Hence the equation $f(\delta_n) = 0$ has a solut-
ion δ_n with $|\delta_n| \leq A \ (c/2)^{2n}$.

One also checks that for this value of δ_n the function R_{n+1} is
really in $L_{2,\gamma_{n+1}}$, if K and K' are sufficiently small. The corresponding condition turns out to be

$$(1 + \delta_n) \ \gamma_{n+1} \ - \ \delta_n \quad \geq \quad \gamma_n \ ,$$

and this is satisfied by construction. The bound on $\| R_{n+1} \|_{2,\gamma_{n+1}}$
follows now easily. This completes the proof of the theorem.

We interpret now Theorem 14.6, by using our previous estimates.
We then see that Theorem 14.6 says that under the repeated application
of the transformation (14.7) our original function $\phi_\epsilon + f$ converges
to a Gaussian. The change of scale during the finite first n_5 steps
can be easily incorporated. By changing back to the original scale, we

get the crossover result for "zero field, above the critical temperature".

THEOREM 14.7. Let $\phi_\varepsilon + f \notin \mathcal{W}_s$, let $\| f \|_\infty \leq \varepsilon^{330}$, $\varepsilon > 0$ suffic-
iently small, let $\phi_\varepsilon + f > 0$, and suppose it is an even function. Sup-
pose furthermore that f is on the side of negative coefficients for
$e_2 \sim +(2x^2-1) \exp(-\varepsilon\theta x^4)$. Then $\mathcal{N}^n(\phi_\varepsilon + f)$ converges to a δ-function
"like a Gaussian" in the following sense.

(i) For some finite constant K, the limit

$$\lim_{n\to\infty} K^{2^n} (2/c)^{n/2} \int dx \ \exp(-x^2/2) \ \mathcal{N}^n(\phi_\varepsilon + f)(x) \ = C_o$$

exists and is different from zero.

(ii) For all m = 0,1,2,..., the limit

$$\lim_{n\to\infty} (2/c)^{mn} \frac{\int dx \ x^{2m} \exp(-x^2/2) \ \mathcal{N}^n(\phi_\varepsilon + f)(x)}{\int dx \ \exp(-x^2/2) \ \mathcal{N}^n(\phi_\varepsilon + f)(x)} = C_m$$

exists, is different from zero, and one has $C_m = C_1 \cdot (2m-1)!!$.

Remark: Theorem 14.6 implies convergence on the larger space $L_{2,\frac{1}{2}-\delta}$,
for $\delta > 0$. Note also that our whole procedure is in principle a compu-
table algorithm, i.e., the constant C_1 can in principle be determined
from $\phi_\varepsilon + f$.

15. Crossover Part II. B) Two Phase Region

In this section, we prove the limiting behaviour for the case $\beta > \beta_{crit}$. The whole of Section 13 applies as before, and we find ourselves after n_2 steps with a function of the form of (13.23), with the plus sign in front of d,

$$\mathcal{N}^{n_2}(\phi_\varepsilon + f) = \text{const.} \left(\phi_\varepsilon + d \, \varepsilon^{15/16} \, e_2 \right.$$
$$\left. + \sum_{j=3}^{5} O(\varepsilon^{29j/32} \, e_{2j}) + O(\varepsilon^{19/8}) \, \exp(-\varepsilon\theta x^4/2) \right) ,$$

with $1 \le d \le 2$. As in Theorem 14.1, this can be brought to the following form.

COROLLARY 15.1. For some constant n_2 , one has for $b = O(1) > 0$,

$$\mathcal{N}^{n_2}(\phi_\varepsilon + f)(x) = \text{const.} \left(\exp(-a'\varepsilon x^4 + b\varepsilon^{15/16} x^2) \right.$$
$$\left. \cdot(1 + \sum_{j=3}^{5} k_j \varepsilon^{29j/32} x^{2j}) + O(\varepsilon^{19/8}) \exp(-a'\varepsilon x^4/2) \right) .$$

Lemma 14.2 follows now as before, and the corollary corresponding to Corollary 14.3 is

COROLLARY 15.2. For some constant n_3 , one has

$$\mathcal{N}^{n_3}(\phi_\varepsilon + f)(x) = \text{const.} \left(\exp(b_3 \varepsilon^{17/32} x^2 - a_3'' \varepsilon x^4) \right.$$
$$\left. + O(\varepsilon^{21/16}) \exp(-a_3'' \varepsilon x^4/2) \right) , \qquad (15.1)$$

with $1 \le b_3 \le 2$, $a_3 = O(1)$.

From this point onwards, the discussion is different from the "high-temperature" case treated in Section 14. Using the perturbation ideas mentioned in Section 6, we get the

PROPOSITION 15.3. For some constant n_4 , one has

$$\mathcal{N}^{n_4}(\phi_\varepsilon + f)(x) \;=\; \text{const.} \left(\exp(b_4 \varepsilon^{\frac{1}{2}} x^2 - a_4 \varepsilon x^4) \right.$$

$$\left. + \; O(\varepsilon^{11/8}) \exp(-a_4 \varepsilon x^4/2) \right) ,$$

where $a_4/8 \leq b_4^2 \leq a_4/4$, $b_4 > 0$, $a_4 = \theta + O(\varepsilon)$, cf. Eq.(3.10) for the definition of θ.

Proof: If $\mathcal{N}^n(\phi_\varepsilon + f)$ is of the form of Eq.(15.1), then every applic-ation of $\mathcal{N}$ multiplies b_3 by $(2/c)+O(\varepsilon)$ and a_3'' by $(2/c^2)+O(\varepsilon)$ and the remainder by $2+O(\varepsilon)$, as we have seen with some variations before. The main observation is that the coefficient A of the quadratic term in the exponential is still sufficiently small with respect to the co-efficient B of the quartic term, so that for all values of n , $n_3 \leq n \leq n_4$, one has

$$\exp(-B x^4 + A x^2) \;\leq\; \exp(-B x^4/2) \, O(1) . \tag{15.2}$$

As before, we absorb then the quartic and quadratic terms into the expo-nential and we dominate the higher order terms by their supremum. This should be easy to reproduce for the reader, to whom we leave further details of the proof of this proposition.

We rewrite $\mathcal{N}^{n_4}(\phi_\varepsilon + f)$ in a different form as

$$\mathcal{N}^{n_4}(\phi_\varepsilon + f)(x) \;=$$

$$= \text{const.} \left(\exp(-\alpha_o \varepsilon x^4 + a_o' x^2 - a_o'^2/(4\alpha_o \varepsilon)) \right.$$

$$\left. + \sum_{\pm} \exp(-a_o (m_o \pm x)^2) \sigma_o(m_o \pm x) + R_o(x) \right) , \tag{15.3}$$

with the following definitions:

$$a_o = A_1 \varepsilon^{\frac{1}{2}} , \quad A_1 = O(1) , \quad a_o = 2a_o' ,$$

$$\alpha_o = O(1) , \quad m_o = (a_o'/2\alpha_o)^{\frac{1}{2}} \varepsilon^{-\frac{1}{2}} = O(\varepsilon^{-\frac{1}{4}}) ,$$

$$\|\sigma_o\|_{\lambda_o,\gamma} = O(\varepsilon^{11/8}), \quad \lambda_o = 4\gamma/(7a_o') =: \varepsilon^{-\frac{1}{2}} L_o + (1-c)/(1-c/2),$$

$$R_o = 0 .$$

The next proposition states that this form is preserved under the repeated action of $\mathcal{N}$.

PROPOSITION 15.4. The function $\mathcal{N}^{n_4+n}(\phi_\varepsilon + f)$ is proportional to a function of the form (15.3) (with parameters α_n, a_n etc.) as long as $n \geq 1$ and $a_{n-1}' \leq \varepsilon^{\frac{1}{4}}(c/2)$. One has the following bounds and recursion relations:

$$\alpha_n = 2c^{-2} \alpha_{n-1} + O(\varepsilon) ,$$

$$a_n' = (2/c) a_{n-1}' + \varepsilon(12\alpha_{n-1}/(c-2a_{n-1}'c)) + O(\varepsilon^2) ,$$

$$m_n = c^{n/2} m_o , \quad a_n = (2/c)^n a_o ,$$

$$\lambda_n = (c/2)^n L_o \varepsilon^{-\frac{1}{2}} + (1-c)/(1-c/2) .$$

Furthermore, if $2^n \leq \varepsilon^{-1/50}$, then

$$\|\sigma_n\|_{\lambda_n,\gamma} \leq (6 + O(\varepsilon^{\frac{1}{4}}))^n O(\varepsilon^{11/8}) , \quad \text{and} \quad R_n(x) = 0 .$$

But if n is such that $\varepsilon^{-1/50} \le 2^n \le \varepsilon^{-\frac{1}{2}}$, then

$$
\| \sigma_n \|_{\lambda_n, \gamma} \quad \le \quad \begin{cases} O(\varepsilon^{21/16}) \ (2 + \varepsilon^{1/6})^{n-n_o'}, & \underline{\text{if }} 2^n \le \varepsilon^{1/25-3/8} \quad , \\ O(\varepsilon^{3/2} \ 2^{3n/2}) & , \ \underline{\text{if }} 2^n > \varepsilon^{1/25-3/8} \quad ; \end{cases}
$$

and

$$
|R_n(x)| \quad \le \quad \exp(\varepsilon^{1/50} \ 2^n \ \log \varepsilon) \ \exp(-b_n m_n^2 \ \rho_n/2 - b_n x^2) \quad ,
$$

with

$$
b_n \quad = \quad a_n - \gamma/\lambda_n \quad , \quad \rho_n \quad = \quad \rho_{n-1} - 2b_{n-1}/(1+2b_{n-1}) ,
$$

and

$$
\rho_{n_o'} \quad = \quad 2/(1 + 2b_{n_o'}) \quad ,
$$

where n_o' is defined by $\varepsilon^{-1/50}/2 < 2^{n_o'} \le \varepsilon^{-1/50}$.

The point of these estimates is that the error terms σ_n and R_n are essentially Gaussians centered at the maxima of the main part and at the origin respectively.

Proof: We proceed by induction from n to n+1. We have

$$
\mathcal{N}^{n_4+n+1}(\phi_\varepsilon +f)(x) \quad = \quad \exp(-2c^{-2} \alpha_n \varepsilon x^4 + 2c^{-1}a_n'x^2 - a_n'^2/(2\alpha_n\varepsilon))
$$

$$
\cdot \ \pi^{-\frac{1}{2}} \int du \ \exp(-u^2(1-2a_n'+12\alpha_n \varepsilon x^2/c)) \ \exp(-2\alpha_n \varepsilon u^4) \quad + \quad R \ .
$$

$$\tag{15.4}$$

We control the integral in (15.4) by expanding

$$
\exp(-2\alpha_n \varepsilon u^4) \quad = \quad 1 - 2\alpha_n \varepsilon u^4 + O(\varepsilon^2 u^8) \quad ,
$$

and integrating over u. This yields for the integral, after expanding the denominators,

$$
(1 - 2a_n')^{-\frac{1}{2}} \left[1 - \rho \varepsilon x^2/2 + 3\rho^2\varepsilon^2 x^4/8 - 3\alpha_n \varepsilon/2(1-2a_n') + \right.
$$

$$+ 15\alpha_n \rho \, \varepsilon^2 x^2/4(1-2a_n')^2 - 105 \, \alpha_n \rho^2 \varepsilon^3 x^4/(1-2a_n')^2$$

$$+ O(\varepsilon^3 x^6) \Bigg\} + O_{L_\infty}(\varepsilon^2) \,, \qquad (15.5)$$

where $\rho = 12\alpha_n/(c-2a_n'c)$. By the recursive relations for the α_n and the a_n', we see that the principal term of the function $\mathcal{N}^{n_4+n+1}(\phi_\varepsilon + f)$ behaves as is claimed in the proposition (up to a factor which is $K_n = 1 + O((2/c)^n a_0')$). We have neglected a term of the form

$$\exp(-\alpha_{n+1}\,\varepsilon\, x^4 + 2c^{-1}a_n' x^2 - a_n'^2/2\alpha_n\varepsilon) \left(O_{L_\infty}(\varepsilon^2) + O(\varepsilon^3 x^6 + \varepsilon^4 x^8)\right),$$

$$(15.6)$$

and this term will now be absorbed into σ_{n+1}, and will thus give a first contribution $\sigma_{n+1}^{(1)}$ to this term. By the symmetry of the problem it suffices to bound

$$\sigma_{n+1}^{(1)}(x + m_{n+1}) = K_n^{-1} \exp(a_{n+1}(x + m_{n+1})^2) \qquad (15.7)$$

$$\cdot \exp(-\alpha_{n+1}\,\varepsilon\,x^4 + 2c^{-1}a_n' x^2 - a_n'^2/2\alpha_n\varepsilon)\,\varepsilon^3\,x^6\,\Theta(-x)\,,$$

where $\Theta(x)$ is the step function. We have to bound the $L_{\lambda_{n+1},\gamma}$ norm of $\sigma_{n+1}^{(1)}$. By the Schwarz inequality, we find

$$\int_{-\infty}^{0} dx \, |\sigma_{n+1}^{(1)}(x)|^{\lambda_{n+1}} \exp(-\gamma(x + m_{n+1})^2)$$

$$\leq \varepsilon^{3\lambda_{n+1}}\, O(1)\, \left(\int_{-\infty}^{0} dx\,|x^{12\lambda_{n+1}}|\exp(-\gamma(x + m_{n+1})^2)\right)^{\frac{1}{2}}$$

$$\leq \varepsilon^{3\lambda_{n+1}}\, O(1)\, (m_{n+1})^{6\lambda_{n+1}}\,,$$

and therefore

$$\| \sigma_{n+1}^{(1)} \|_{\lambda_{n+1},\gamma} \leq O(1)\, \varepsilon^3\, m_{n+1}^6\,.$$

We next treat the terms which contribute to the remainder R. For this
we rewrite first the principal part of (15.3) (i.e. its first term) as

$$\sum_{\pm} \sum_n (m_n \pm x) \, \exp(\, -a_n \, (m_n \pm x)^2 \,) \quad,$$

with $\operatorname{supp}\Sigma_n \subset [-\infty, \, m_n]$, and we estimate $\| \Sigma_n \|_{\lambda_n, \gamma}$. This will allow
us to treat the principal term on the same footing as the second and
third term in (15.3). Using the relations between a_n' , α_n , m_n ,and
λ_n one sees that

$$\| \Sigma_n \|_{\lambda_n, \gamma}^{\lambda_n} \quad \le \quad \exp(\, \lambda_n \, O(\varepsilon^{\frac{1}{2}} \, 2^n))$$

$$\cdot \int_{-\infty}^{m_n} dx \, \exp(\, -\alpha_n \, \varepsilon \, x^4 \lambda_n \, + \, 4\lambda_n \alpha_n \, \varepsilon \, m_n x^3 \, - \, (\gamma + O(\varepsilon^{\frac{1}{2}}))x^2 \, + \, O(\varepsilon^{\frac{1}{4}})x \,)$$

$$= \quad \exp(\, \lambda_n \, O(\varepsilon^{\frac{1}{2}} \, 2^n) \,) \quad.$$

We now start with the two remainder terms proper. We shall introduce
a standard numbering for the six typical terms whose sum is R. As we
have just seen, the function (15.3) can also be written as

$$\sum_{\pm} \exp(\, -a_n (m_n \pm x)^2) \quad (\, \Sigma_n (m_n \pm x) \, + \, \sigma_n (m_n \pm x) \,) \quad + \quad R_n$$

$$= \quad A_+ \; + \; a_+ \; + \; A_- \; + \; a_- \; + \; R \; .$$

We view again $\mathcal{N}$ as a bilinear map, and then the contributions to R are:

$$R_1 \quad = \quad \mathcal{N}(a_+, a_+) \; + \; 2 \, \mathcal{N}(A_+, a_+) \quad,$$

$$R_2 \quad = \quad \mathcal{N}(a_-, a_-) \; + \; 2 \, \mathcal{N}(A_-, a_-) \quad,$$

$$R_3 \quad = \quad \mathcal{N}(R, R) \quad,$$

$$R_4 \quad = \quad 2 \, \mathcal{N}(A_+ + a_+, \, R) \quad,$$

$$R_5 \quad = \quad 2 \, \mathcal{N}(A_- + a_-, \, R) \quad,$$

$$R_6 \quad = \quad 2 \, \mathcal{N}(A_+, a_-) \; + \; 2 \, \mathcal{N}(A_-, a_+) \; + \; 2 \, \mathcal{N}(a_+, a_-) \quad.$$

It clearly suffices to deal with R_1, R_3, R_4, R_6.

Control of R_1.

We only deal with the second term, the first one is similar. To bound this term, called R_{11}, we write

$$|R_{11}(x)| \leq 2\pi^{-\frac{1}{2}} \int du \exp(-u^2 - a_n(xc^{-\frac{1}{2}}-u+m_n)^2 - a_n(xc^{-\frac{1}{2}}+u+m_n)^2)$$

$$\cdot |\Sigma_n(xc^{-\frac{1}{2}}-u+m_n) \ \sigma_n(xc^{-\frac{1}{2}}+u+m_n)| \quad .$$

We want to absorb this term into $\exp(-a_{n+1}(x+m_{n+1})^2) \ \sigma_{n+1}^{(2)}(x+m_{n+1})$,

with $\sigma_{n+1}^{(2)} \in L_{\lambda_{n+1},\gamma}$. Thus we have to bound

$$\pi^{-\frac{1}{2}} \gamma^{\frac{1}{2}} \int dt \exp(-\gamma t^2) \qquad (15.7)$$

$$\cdot \left[\pi^{-\frac{1}{2}} \int du \exp(-(1+2a_n)u^2) \ |\Sigma_n(tc^{-\frac{1}{2}}+u) \ \sigma_n(tc^{-\frac{1}{2}}-u)| \right]^{\lambda_{n+1}} \quad .$$

By the Schwarz inequality, the square bracket is bounded by

$$(\pi^{-\frac{1}{2}} \int du \exp(-u^2) \ |\Sigma_n(tc^{-\frac{1}{2}}-u)|^2)^{\frac{1}{2}}$$

$$\cdot (\pi^{-\frac{1}{2}} \int du \exp(-u^2) \ |\sigma_n(tc^{-\frac{1}{2}}-u)|^2)^{\frac{1}{2}}$$

$$= \frac{1}{4} (\mathcal{A}_1(|\Sigma_n|^2)(t))^{\frac{1}{2}} \cdot (\mathcal{A}_1(|\sigma_n|^2)(t))^{\frac{1}{2}} \quad .$$

But by Lemma 10.7 (hypercontractivity), and using again the Schwarz inequality, we find that (15.7) is bounded by

$$(1/4)^{\lambda_{n+1}} \ \| \mathcal{A}_1(|\Sigma_n|^2) \|_{\lambda_{n+1},\gamma}^{\lambda_{n+1}} \cdot \| \mathcal{A}_1(|\sigma_n|^2) \|_{\lambda_{n+1},\gamma}^{\lambda_{n+1}}$$

$$\leq \left\{ \| \Sigma_n \|_{2((\lambda_{n+1}+1)/c - 1),\gamma} \cdot \| \sigma_n \|_{2((\lambda_{n+1}+1)/c - 1),\gamma} \right\}^{\lambda_{n+1}} ,$$

so that this contribution yields

$$\| \sigma_{n+1}^{(2)} \|_{\lambda_{n+1}, \gamma} \leq (2(1+O(\varepsilon^{\frac{1}{2}}/\lambda_n)) \exp(O(\varepsilon^{\frac{1}{2}}2^n)) + \| \sigma_n \|_{\lambda_n, \gamma})$$

$$\cdot \| \sigma_n \|_{\lambda_n, \gamma} \cdot$$

The other term of R_1 gives the same bound.

Control of R_3.

This is easy. If $|R_n(x)| \leq r_n \exp(-b_n x^2)$, then $|\mathcal{N}(R_n)(x)| \leq$
$r_n^2 (1 + 2b_n)^{-\frac{1}{2}} \exp(-2c^{-1}b_n x^2)$, as can be seen by explicitly integrating the bound.

Control of R_4.

Assume again that $|R_n(x)| \leq r_n \exp(-b_n x^2)$, and let $\sigma_n' = \Sigma_n + \sigma_n$. Then $|R_4(x)|$ is bounded by

$$2\pi^{-\frac{1}{2}} r_n \int du \exp\left(-u^2 - a_n (xc^{-\frac{1}{2}} + m_n + u)^2 - b_n (xc^{-\frac{1}{2}} - u)^2\right) |\sigma_n'(xc^{-\frac{1}{2}} + m_n + u)| \cdot$$

$$(15.8)$$

By the Hölder inequality, this is in turn bounded by

$$2\pi^{-\frac{1}{2}} r_n \| \sigma_n' \|_{\lambda_n, \gamma} \exp\left(-2c^{-1}b_n(x+m_nc^{\frac{1}{2}}/2)^2 - b_n m_n^2/2 + b_n^2 m_n^2/(1+2b_n)\right)$$

$$\cdot \left(\pi^{\frac{1}{2}}/(q(1+2b_n))\right)^{1/q} (\pi/\gamma)^{1/2\lambda_n} \tag{15.8'}$$

$$\leq 2r_n (1 + O(\varepsilon^{1/6})) \exp\left(-2c^{-1}b_n(x+m_nc^{\frac{1}{2}}/2)^2 - b_n m_n^2/(2+4b_n)\right) \cdot$$

We have used $q^{-1} + \lambda_n^{-1} = 1$, and $b_n = a_n - \gamma/\lambda_n$. We shall now absorb this bound into σ_{n+1} and into R_{n+1}. For this we split off the x-dependent Gaussian factors in (15.8) and write it as

$$|R_4(x)| \leq 2 \pi^{-\frac{1}{2}} r_n (\pi\gamma^{-1})^{1/2\lambda_n} \tag{15.9}$$

$$\cdot \exp\left(-2c^{-1}b_n(x+m_nc^{\frac{1}{2}})^2 + 2b_n c^{-\frac{1}{2}}m_n x + b_n m_n^2\right) \cdot H(x) ,$$

and we define x_n by

$$r_n \, \varepsilon^{-3/2} \exp\left(-2c_n^{-1}b_n(x_n+m_nc^{\frac{1}{2}}/2)^2 - b_nm_n^2/(2+4b_n)\right)$$

$$= \exp\left(-2b_nc_n^{-1}(x_n + m_nc^{\frac{1}{2}})^2\right) \quad . \tag{15.10}$$

Note that $x_n < 0$. For $x < x_n$ we absorb the bound (15.8) into

$\sigma_{n+1}(m_{n+1}+x)$, while for $x \geq x_n$ we absorb it into $R_{n+1}(x)$. Indeed in

the case $x < x_n$ we have by (15.10),

$$\exp\left(2b_nc^{-\frac{1}{2}}m_nx\right) \leq \exp\left(2b_nc^{-\frac{1}{2}}m_nx_n\right)$$

$$= r_n^{-1} \varepsilon^{3/2} \exp\left(-3b_nm_n^2/2 + b_nm_n^2/(2+4b_n)\right) \quad .$$

Therefore, we have in this case, by (15.9),

$$|R_4(x)| \leq O(\varepsilon^{3/2}) \, H(x) \, \exp\left(-2c_n^{-1}b_n(x+m_nc^{\frac{1}{2}})^2\right)$$

$$\cdot \exp\left(-b_nm_n^2/2 + b_nm_n^2/(2+4b_n)\right)$$

$$= O(\varepsilon^{3/2}) \, H(x) \, \exp\left(-2c_n^{-1}b_n(x+m_nc^{\frac{1}{2}})^2 - b_n^2m_n^2/(1+2b_n)\right) \quad .$$

Using the Hölder inequality, one checks easily that

$$\| R_4(x) \, \Theta(x_n-x) \, \exp\left((a_{n+1}-\gamma/\lambda_{n+1})(x+m_{n+1})^2\right) \|_{\lambda_{n+1},\gamma} \leq O(\varepsilon^{3/2}) \quad .$$

In the case $x \geq x_n$, we get from (15.8') the bound

$$|R_4(x)| \leq O(\varepsilon^{-3/2}) \, r_n^2 \, \exp\left(2b_n^2m_n^2/(1+2b_n) - 2b_nc^{-1}x^2\right) \quad .$$

Control of R_6.

There are two different methods according to whether $2^n \leq \varepsilon^{-1/50}$ or

$2^n > \varepsilon^{-1/50}$. We first deal with the first case. In this case the contribution will be absorbed into $\sigma_{n+1}(m_{n+1}+x)+\sigma_{n+1}(m_{n+1}-x)$. We deal only with $x < 0$ which is absorbed into $\sigma_{n+1}(m_{n+1}+x)$. The corresponding contribution is typically of the form

$$2\pi^{-\frac{1}{2}} \ \exp\left((-a_{n+1} - \gamma/\lambda_{n+1})(m_{n+1}+x)^2\right) \ (\gamma/\pi)^{1/\lambda_n}$$

$$\cdot \exp\left(2c^{-1}(a_n - \gamma/\lambda_n) m_{n+1}^2\right)$$

$$\cdot \int du \ \exp\left(-u^2 - 2(a_n - \gamma/\lambda_n)(m_n+u)^2\right) \ |\sigma'(xc^{-\frac{1}{2}}+m_n+u) \cdot \sigma''(-xc^{-\frac{1}{2}}+m_n+u)| ,$$

where σ', σ'' equal Σ_n or σ_n (but not both are equal to Σ_n). By the methods used to bound R_4, one finds that the contribution $\sigma_{n+1}^{(3)}$ to σ_{n+1} is bounded by

$$(1 + O(\varepsilon^{1/6})) \ (\ 4\ ||\sigma_n||_{\lambda_n,\gamma} + 2\ ||\sigma_n||^2_{\lambda_n,\gamma}\ ) \ \exp(4b_n^2 m_n^2/(1+2b_n))$$

$$\leq \ (\ 4 + O(\varepsilon^{1/6})\ )\ ||\sigma_n||_{\lambda_n,\gamma} \ , \tag{15.11}$$

since for $2^n \leq \varepsilon^{-1/50}$ the exponential in (15.11) is bounded by $(1 + O(\varepsilon^{1/6}))$.

In the __second case__ we cannot use this last fact any more and we absorb the contribution into R_{n+1}. A computation as in case R_4, using the Hölder inequality, yields (with $p^{-1} + 2\lambda_n^{-1} = 1$),

$$|R_6(x)| \ \leq \ \pi^{-\frac{1}{2}} \ (\pi/\gamma)^{1/\lambda_n} \ \left((\ ||\Sigma_n||_{\lambda_n,\gamma} + ||\sigma_n||_{\lambda_n,\gamma})^2 - ||\Sigma_n||^2_{\lambda_n,\gamma}\right)$$

$$\cdot \exp\left(-(a_n-\gamma/\lambda_n)(2c^{-1}x^2 + m_n^2/(1 + 2a_n - 2\gamma/\lambda_n))\right)$$

$$\cdot (\ \pi/(p(1 + 2a_n - 2\gamma/\lambda_n)))^{1/2p}$$

$$\leq \ (1+O(\varepsilon^{1/6})) \ 2||\sigma_n||_{\lambda_n,\gamma} \ \exp\left(-b_{n+1}x^2 - (a_n-\gamma/\lambda_n)m_n^2/(1+2a_n-2\gamma/\lambda_n)\right).$$

We now <u>summarize the estimates</u>: For $2^n \leq \varepsilon^{-1/50}$, we have shown the relations

$$r_{n+1} \quad = \quad 0 \quad ,$$

$$\| \sigma_{n+1} \|_{\lambda_{n+1}, \gamma} \leq \varepsilon^3 m_{n+1}^6 + 2(1 + O(\varepsilon^{1/6})) \exp(O(\varepsilon^{\frac{1}{2}} 2^n)) \| \sigma_n \|_{\lambda_n, \gamma}$$

$$+ \quad 4(1 + O(\varepsilon^{1/6})) \| \sigma_n \|_{\lambda_n, \gamma} \quad .$$

In the case $2^n > \varepsilon^{-1/50}$, we have shown

$$r_{n+1} \quad \leq \quad r_n^2 (1 + O(\varepsilon^{-3/2}) \exp(2 b_n^2 m_n^2 / (1 + 2 b_n)))$$

$$+ \quad 2 (1 + O(\varepsilon^{1/48})) \| \sigma_n \|_{\lambda_n, \gamma} \exp(-b_n m_n^2 / (1 + 2 b_n)),$$

$$\| \sigma_{n+1} \|_{\lambda_{n+1}, \gamma} \leq \varepsilon^3 m_{n+1}^6 + 2(1 + O(\varepsilon^{1/48})) \exp(O(\varepsilon^{\frac{1}{2}} 2^n)) \| \sigma_n \|_{\lambda_n, \gamma}$$

$$+ \quad \varepsilon^{3/2} \quad . \tag{15.12}$$

The proof of the proposition is then accomplished by checking that these relations are compatible with the recursive bounds stated in the proposition. As an immediate consequence of the Proposition 15.4 we get the

<u>COROLLARY 15.5.</u> <u>There is a number</u> $n_5 > n_4$ <u>such that one has with</u> $n = n_5 - n_4$ (= $O(\log \varepsilon^{-1})$) , <u>the relation</u>

$$\mathcal{N}^{n_5}(\phi_\varepsilon + f)(x) \quad = \quad \text{const.} \left(\exp(-\alpha_n \varepsilon x^4 + a_n' x^2 - a_n'^2 / (4 \alpha_n \varepsilon)) \right.$$

$$+ \sum_{\pm} \exp(-a_n (m_n \pm x)^2) \cdot \sigma_n (m_n \pm x) \quad + \quad R_n(x) \left. \right) , \tag{15.13}$$

<u>with</u> $\varepsilon^{\frac{1}{2}}/2 < a_n' \leq \varepsilon^{\frac{1}{2}}$, <u>and where all other bounds are as in Proposit-</u>
<u>ion 15.4.</u>

We write this now in a double Gaussian form.

LEMMA 15.6. There is a number $n_5 > n_4$ such that with $n = n_5 - n_4$, the function $\mathcal{N}^{n_5}(\phi_\varepsilon + f)$ can be written in the form

$$\mathcal{N}^{n_5}(\phi_\varepsilon + f)(x) \qquad\qquad (15.14)$$

$$= \text{const.} \left[\sum_{\pm} \exp(-\hat{a}_n(\hat{m}_n \pm x)^2) \left(1 + P_n(\hat{m}_n \pm x) + \hat{\sigma}_n(\hat{m}_n \pm x)\right) + R_n(x) \right],$$

with the following definitions and relations:

$$P_n(x) = O(\varepsilon^{3/8}) \; H_3(\gamma^{\frac{1}{2}}x) + O(\varepsilon^{5/4}) \; H_6(\gamma^{\frac{1}{2}}x) + O(\varepsilon) \; H_4(\gamma^{\frac{1}{2}}x) \quad,$$

$$\hat{a}_n = 2a_n' + O(\varepsilon^{3/8}) = a_n + O(\varepsilon^{3/8}) \quad, \quad \hat{m}_n = m_n + O(\varepsilon^{1/8}) \quad,$$

$$\|\hat{\sigma}_n\|_{\hat{\lambda}_n,\gamma} \leq O(\varepsilon^{3/4}) \quad, \quad \hat{\lambda}_n = \lambda_n(1 - \varepsilon^{\frac{1}{2}}/\gamma).$$

Proof: This is a rearrangement of the formula (15.13). We first handle the principal term. Let $y = m_n + x$. Then the principal term is

$$\exp(-\alpha_n \varepsilon x^4 + a_n' x^2 - a_n'^2/(4\alpha_n \varepsilon))$$

$$= K \exp(-\alpha_n \varepsilon y^4 + 4\alpha_n m_n \varepsilon y^3 - a_n y^2 + sy^2 + ty)$$

and the combination of the recursive relations for $a_n,\ \alpha_n$,... with Eq. (15.5) shows that

$$s = O(\varepsilon \log \varepsilon) \quad, \qquad t = O(\varepsilon^{5/8} \log \varepsilon) \quad, \quad K = O(1) \quad.$$

Concentrating now on the side of negative x, we consider

$$G(y) = \exp(-a_n y^2) \left[\exp\left(-\alpha_n \varepsilon y^4 + 4\alpha_n m_n \varepsilon y^3 + sy^2 + ty\right) + K^{-1}\sigma_n(y) \right] \quad.$$

We substitute $y = z + \delta$, and we look for a δ (near zero) such that G takes the form

$$G(y) = O(1) \exp(-\hat{a}_n z^2) \tag{15.15}$$

$$\cdot \left[\exp\left(-\alpha_n \varepsilon z^4 + (4\alpha_n m_n \varepsilon - 4\alpha_n \delta)(z^3 - 3z/4\gamma)\right) + \tilde{\sigma}_n(z) \right] .$$

This condition leads in fact to the equation

$$-4\alpha_n \varepsilon \delta^3 + 3\delta^2 4\alpha_n m_n \varepsilon + \delta(2a_n - 2s + 3\alpha_n/\gamma) + t + 3\alpha_n m_n \varepsilon/\gamma = 0.$$

This equation has, due to the bounds on the various coefficients appearing in it, a unique solution in the interval $|\delta| \le O(-\varepsilon^{3/8} \log \varepsilon)$. As a consequence

$$\tilde{\sigma}_n(z) = O(1) \sigma_n(z+\delta) \exp\left(O(\varepsilon \log \varepsilon) z^2 + O(\varepsilon^{5/8} \log \varepsilon) z\right) .$$

The only remaining difficulty is to show that $\tilde{\sigma}_n$ is appropriately bounded. We achieve this by slightly reducing λ_n in order to compensate the (unbounded) translation by δ in $L_{\lambda_n,\gamma}$. In fact, we choose $\hat{\lambda}_n = \lambda_n(1-\varepsilon^{\frac{1}{2}}/\gamma)$. Then one can check that $\| \tilde{\sigma}_n \|_{\hat{\lambda}_n,\gamma} \le O(\varepsilon^{3/4})$.

To complete the proof, one expands the exponential factor in Eq. (15.15) in the way we have done several times before, see e.g. the case of R_1 in Proposition 15.4. We absorb higher order terms in $\tilde{\sigma}_n$ and leave the lower order terms as they are and they will then form the function P_n.

Note: The proof of this lemma implied several small changes of order $O(\varepsilon^\alpha)$, $\alpha > 0$ for our recursively defined constants. We have denoted the new quantities with a " ^ " . Starting from $n > n_5$, however, we shall omit the hat " ^ " again, so that, e.g., the recursion relation for $a_{n+n_5-n_4}$ is for $n > 0$

$$a_{n+n_5-n_4} = (2/c)^n \, \hat{a}_{n_5} = (2/c)^n \left(a_o (2/c)^{n_5-n_4} + O(\varepsilon^{3/8})\right) .$$

We next state a variant of Proposition 15.4 for the case
$n + n_4 > n_5$.

LEMMA 15.7. For $n + n_4 > n_5$ the function $\phi_n = \mathcal{N}^{n_4+n}(\phi_\varepsilon + f)$ is of
the form

$$\phi_n(x) = \text{const.} \left(\sum_\pm \exp(-a_n(m_n \pm x)^2) \, (1 + P_n(m_n \pm x) + \sigma_n(m_n \pm x)) \right.$$

$$\left. + R_n(x) \right) ,$$

with the recursive definitions and bounds:

$$a_n = (2/c)^n \, \hat{a}_{n_5-n_4} \quad , \quad m_n = c^{n/2} \, \hat{m}_{n_5-n_4} \quad ,$$

$$P_n(x) = \alpha_n H_3(\gamma^{\frac12}x) + \beta_n H_4(\gamma^{\frac12}x) + \rho_n H_6(\gamma^{\frac12}x) \quad ,$$

$$\alpha_{n+1} = 2c^{-3/2}\alpha_n \quad , \quad \beta_{n+1} = 2c^{-2}\beta_n \quad , \quad \rho_{n+1} = 2c^{-3}\rho_n + \alpha_n^2 \quad ,$$

$$\alpha_{n_5-n_4} = O(\varepsilon^{3/8}) \quad , \quad \beta_{n_5-n_4} = O(\varepsilon) \quad , \quad \rho_{n_5-n_4} = O(\varepsilon^{5/4}) \quad ,$$

$$\| \sigma_{n+1} \|_{\lambda_{n+1},\gamma} = O(\varepsilon^{3/4}) + (2 + O(\varepsilon^{1/48})) \| \sigma_n \|_{\lambda_n,\gamma} + \varepsilon^{3/2} \quad .$$

The number λ_{n+1} depends in the same way on λ_n as in Proposition 15.4.
Finally, $|R_n(x)| \le r_n \exp(-b_n x^2)$, with the recurrence relations
(15.12). All these estimates are valid as long as $a_n \le \varepsilon^{1/24}$.

Proof: The proof is identical to the one of Proposition 15.4 except for
the principal term. A typical contribution in the iteration from n to
$n+1$ is

$$(1 + 2a_n)^{-\frac12} \exp(-a_{n+1}(m_{n+1} + x)^2) \quad + \qquad\qquad (15.16)$$

$$+ 2\pi^{-\frac{1}{2}} \exp(-a_{n+1}(m_{n+1}+x)^2)$$

$$\cdot \int du \exp(-u^2(1+2a_n)) \ \left(P_n(tc^{-\frac{1}{2}}-u) + \sigma_n(tc^{-\frac{1}{2}}-u)\right)$$

$$+ \ \pi^{-\frac{1}{2}} \exp(-a_{n+1}(m_{n+1}+x)^2) \ \int du \exp(-u^2(1+2a_n))$$

$$\cdot \left(P_n(tc^{-\frac{1}{2}}-u) + \sigma_n(tc^{-\frac{1}{2}}-u)\right) \cdot \left(P_n(tc^{-\frac{1}{2}}+u) + \sigma_n(tc^{-\frac{1}{2}}+u)\right) \quad ,$$

where $t = m_{n+1} + x$. The only new type of terms in (15.16) are the ones involving the polynomials P_n. They are bounded by perturbation theory, as we have done several times in Section 13 (cf. e.g. Eq.(13.9)). Then the assertion of the lemma follows at once.

COROLLARY 15.8. There is a number $n_6 > n_5$ such that with $n = n_6 - n_4$, one has

$$\mathscr{N}^{n_6}(\phi_\varepsilon + f)(x) = \text{const.} \ \left[\ \sum_{\pm} \exp(-a_n(m_n \pm x)^2) \ (1 + \sigma_n'(m_n \pm x)) \right.$$

$$\left. + R_n(x) \ \right] \quad ,$$

where $\varepsilon^{1/24}/2 \le a_n \le \varepsilon^{1/24}$, and one has $\| \sigma_n' \|_{\lambda_n, \gamma} \le O(\varepsilon^{5/24})$, and R_n is bounded as before.

Proof: We simply absorb P_n into σ_n' , since $\| P_n \|_{\lambda_n, \gamma} = O(\varepsilon^{5/24})$, where one uses that $\lambda_n \sim O(\varepsilon^{-1/24})$, and that the coefficient of H_3 is $\varepsilon^{3/8}$ $\cdot \varepsilon^{1/48 - 1/8}$.

Note: The note following the proof of Lemma 15.6 applies now, mutatis mutandis, to the quantity σ_n' , i.e. for $n > n_6 - n_4$ the recurrence relations are based on $\sigma_{n_6 - n_4}'$ rather than $\sigma_{n_6 - n_4}$, but we shall omit henceforth the prime.

We next state a lemma without proof, since it is essentially identical to the proofs of Proposition 15.4 and Lemma 15.7.

LEMMA 15.9. For $n+n_4 > n_6$ the function $\phi_n = \mathcal{N}^{n_4+n}(\phi_\varepsilon + f)$ is of the form

$$\phi_n(x) = \text{const.} \left[\sum_{\pm} \exp(-a_n(m_n \pm x)^2)(1 + \sigma_n(m_n \pm x)) + R_n(x) \right] ,$$

with the following recursive bounds:

$$a_n = (2/c) a_{n-1} , \qquad m_n = c^{\frac{1}{2}} m_{n-1} ,$$

$$\| \sigma_n \|_{\lambda_n,\gamma} \leq \varepsilon^{3/2} + \| \sigma_{n-1} \|^2_{\lambda_{n-1},\gamma}$$

$$+ 2 \| \sigma_{n-1} \|_{\lambda_{n-1},\gamma} \left((1 + 2a_{n-1})/(1 + 2b_{n-1}) \right)^{\frac{1}{2}} ,$$

$$|R_n(x)| \leq r_n \exp(-b_n x^2) ,$$

where b_n is defined in Proposition 15.4, and

$$r_n \leq O(\varepsilon^{-3/2}) r_{n-1}^2 \exp \left(2b_{n-1}^2 m_{n-1}^2/(1+2b_{n-1}) \right)$$

$$+ O(1) \| \sigma_{n-1} \|_{\lambda_{n-1},\gamma} \exp(-b_{n-1}m_{n-1}^2/(1+2b_{n-1})) .$$

The initial values (for $n = n_6-n_4$) are those reached in Corollary 15.8. The estimates above are valid as long as $a_n \leq 0{,}018$.

We fix n_7 in such a way that $0{,}01271 < a_{n_7-n_4} \leq 0{,}018$. We have now reached in our proof the point where the inverse of the covariance of the two Gaussians is macroscopic, and before proceeding further, we want to comment on the situation.

We see now that the function $\mathcal{N}^{n_7}(\phi_\varepsilon + f)$ has a form which is essentially independent of f and of $\varepsilon > 0$, while the number n_7 does depend on these quantities. This reflects the fact that the volume (i.e. 2^{n_7}) for which two different systems have similar spin distributions does depend on the details of the two systems, while the spin distribution itself is (up to a scale factor) essentially independent of the details of these systems. In a sense, we could therefore complete the proof of the crossover by just doing a numerical calculation for the fixed distribution we have obtained so far after n_7 steps. However, while such considerations have been helpful for the construction of the proof, they are not sufficiently precise in order to make sure that the (small) corrections to this universal behaviour (the terms R and σ) do not matter after all.

We should also comment on the separation of the phases, which will take place in the next few steps after the n_7 steps we have done so far. Recall that our final aim is not to consider the function $\mathcal{N}^{n_7}(\phi_\varepsilon + f)$, but the **measure** proportional to $dx \, \exp(-x^2/2) \, \mathcal{N}^{n_7}(\phi_\varepsilon + f)(x)$, and the question of interest is therefore whether this measure will split into two Gaussians. Again a numerical calculation with the essentially universal measure obtained after n_7 steps, or a more heuristic manual computation which does not take into account the subtle effects of R and σ show that the two distinct Gaussians form within only 2 steps. This change will in fact take place for that value of n for which $a_n \sim 1$ (after the first step). Intuitively, the argument is as follows: When $a_n \sim 1$, then

$$\mathcal{N}^n(\phi_\varepsilon + f)(x) \sim \exp\left(-(m_n + x)^2 - (m_n - x)^2\right)$$

$$\sim \exp\left(x^2/2 - \varepsilon\theta x^4\right) \quad , \text{ with } m_n^2 = 1/(4\varepsilon\theta) .$$

The factor $\exp(-x^2/2)$ of the measure kills the factor $\exp(x^2/2)$ of the function and we remain with the very flat function $\exp(-\varepsilon\theta x^4)$. One step earlier, we must have had $a_n \sim c/2$, and we find essentially one Gaussian, while one step later, $a_n \sim 2/c$, so that we have then essentially two Gaussians. The fact that we will always pass through this stage (since the function has after n_7 steps a standard form) shows that the Figure 8c is indeed very typical (page 66), with $N = 3$ being the "flat" function. Unfortunately, the transition during the two crucial steps is so violent that we shall, in the proof, introduce the measure only when $a_n \sim 200$. This seems more convenient for estimating the errors, but it has the disadvantage that the statements which we prove will not make very transparent the beautiful feature of the sudden formation of the two phases.

Let us now summarize the form of $\mathcal{N}^{n_7}(\phi_\varepsilon + f)$.

PROPOSITION 15.10. For a suitable number n_7 the function $\mathcal{N}^{n_7}(\phi_\varepsilon + f)$ is of the form

$$\mathcal{N}^{n_7}(\phi_\varepsilon + f)(x) = \sum_{\pm} \exp(-a_n(m_n \pm x)^2)(1 + \sigma_n(m_n \pm x)) + R_n(x) ,$$

with $0,01271 < a_n \le 0,018$, where $n = n_7 - n_4$, and with bounds and definitions as given below.

Proof: The proposition is of course just a reformulation of Lemma 15.9, but we shall take some care to explain where the different constants come from. Let $n = n_7 - n_4$.

The magnetization m_n. We claim that $m_n = O(\varepsilon^{-\frac{1}{2}})$. This follows from the fact that m_0 in Proposition 15.4 was $O(\varepsilon^{-\frac{1}{4}})$ (cf. Eq.(15.3)) while a_o was $O(\varepsilon^{\frac{1}{2}})$. Each of the $n = n_7 - n_4$ steps multiplied a_o by $(2/c)$. More precisely, $a_n = (2/c)^{n-n_5+n_4} \left(a_o (2/c)^{n_5-n_4} + O(\varepsilon^{3/8})\right)$, as follows from the "note" after Lemma 15.6. Since $a_o (2/c)^{n_5-n_4} = O(\varepsilon^{\frac{1}{4}})$ by Corollary 15.5, we find that n satisfies $(2/c)^n = O(\varepsilon^{-\frac{1}{2}})$ and hence $m_n = c^{n/2} m_o = O(\varepsilon^{-\frac{1}{4}})$, since $c = 2^{\frac{1}{2}} - O(\varepsilon)$.

The error term σ_n. We claim that $\| \sigma_n \|_{17,\gamma} \le O(\varepsilon^{1/8})$. Here the number 17 is the most pessimistic bound on λ_n for the values of a_n we are considering. Indeed, neglecting for the ease of presentation the special handling after step number n_5 , we see that $\lambda_n = 8\gamma/(7a_n) + (1-c)/(1-c/2)$, and this is greater than 17. Therefore $\| \sigma_n \|_{17,\gamma} \le \| \sigma_n \|_{\lambda_n,\gamma}$, and it suffices to bound the latter quantity. In Corollary 15.8 we showed

$$\| \sigma_{n_6-n_4} \|_{\lambda_{n_6-n_4},\gamma} \le O(\varepsilon^{5/24}) ,$$

while $a_{n_6-n_4} = O(\varepsilon^{1/24})$. Then it follows that $(2/c)^{n_7-n_6} = O(\varepsilon^{-1/24})$ and hence the recursion relation of Lemma 15.9 implies

$$\| \sigma_n \|_{\lambda_n,\gamma} \le O(\varepsilon^{5/24}) O(\varepsilon^{-2/24}) ,$$

since the product of the factors $\left((1+2a_{j-1})/(1+2b_{j-1})\right)^{\frac{1}{2}}$ for $n_6-n_4 < j \le n_7 - n_4$ is $O(1)$, as can be checked easily.

The error term R_n. The recurrence relations for R_j are given in Eq. (15.12), for $0 < j \le n_6 - n_4$, and in Lemma 15.9 for $n_6 - n_4 < j < n_7 - n_4$. One can check that the formulas given in Proposition 15.4 remain valid for all these j . Therefore we have

$$|R_n(x)| \le \exp(\varepsilon^{-49/50} \log \varepsilon) \exp(-b_n m_n^2 \rho_n/2) \exp(-b_n x^2/2) ,$$

$$\rho_n = 2/(1 + 2b_{n_o'}) - \sum_{j=n_o'}^{n_7-n_4-1} 2b_j/(1 + 2b_j) ,$$

where

$$b_j = a_j - \gamma/\lambda_j \ , \qquad 2^{n_0'} \sim \varepsilon^{-1/50} \ .$$

For the values of a_n we have chosen, it follows that one has

$0,000951 < b_n < 0,000956$, and $0,99768 < \rho_n/2 < 0,99771$. We shall re-write this bound as

$$|R_n(x)| \leq s_n \exp(-b_n x^2) \ ,$$

with

$$s_n = \exp(-L_n) \ , \quad L_n = \rho_n b_n m_n^2 / 2 \ .$$

Now the stage is set for the 27 more iteration steps during which we expand everything explicitly.

LEMMA 15.11. For $0 \leq j \leq 27$, the function $\mathcal{N}^{j+n_7}(\phi_\varepsilon + f)$ is of the form

$$\text{const.} \left[\sum_{\pm} \exp(-a_n(m_n \pm x)^2)(1 + \sigma_n(m_n \pm x)) + R_n(x) \right] \ ,$$

where $n = j + n_7 - n_4$ and

$$|R_n(x)| \leq s_n \sum_{k=1-2^j}^{2^j-1} \exp(L_n k^2) \ \exp(-b_n(x + km_n 2^{-n})^2) \ .$$

One has the following recursion relations:

$$a_{n+1} = (2/c) a_n \ , \quad m_{n+1} = c^{\frac{1}{2}} m_n \ , \quad b_{n+1} = (2/c) b_n \ ,$$

and as long as $L_n \leq b_n m_n^2/((1 + 2b_n)2^{2j})$, one has

$$s_{n+1} = O(1) s_n^2 \exp(2^{2j+3}) \ , \quad \text{and} \quad L_{n+1} = L_n/2 - 2 \ ,$$

while in the opposite case

$$s_{n+1} = O(1) \ s_n^2 \ \exp(2^{2j+3}) \ \exp(2^{2j+3} \ b_n m_n^2 \ (L_n/2 - b_n m_n^2 / ((2+4b_n) 2^{2j})))$$

and $\ L_{n+1} = b_n m_n^2 / ((2+4b_n) 2^{2j}) - 2$. Finally

$$\| \sigma_{n+1,1} \|_{17,\gamma} \leq 2 (1 + 2a_n)^{\frac{1}{2}} \| \sigma_{n,1} \|_{17,\gamma} \ , \tag{15.17}$$

$$\sigma_n = \sigma_{n,1} + \sigma_{n,2} \quad \text{and}$$

$$|\sigma_{n,1}(z)| \leq O(\| \sigma_{n,1} \|_{17,\gamma}) \ \exp((a_n - b_n) \ z^2) \ , \tag{15.18}$$

$$|\sigma_{n,2}(z)| \leq t_n \ \exp((a_n - b_n) \ z^2) \ , \tag{15.19}$$

with the recursive relation

$$t_{n+1} = O(t_n + \| \sigma_{n,1} \|_{17,\gamma})$$

and the initial values

$$t_{n_7 - n_4} = 0 \ , \quad \| \sigma_{n_7 - n_4, 1} \|_{17,\gamma} \leq O(\varepsilon^{1/8}) \ .$$

Proof: By our previous discussion, the initial hypotheses for $j = 0$ are satisfied. We now control an iteration step. Again, we view $\mathcal{N}$ as a bilinear map and discuss the different terms separately. The first case is that of the dominant part.

$$T_1(x) = \pi^{-\frac{1}{2}} \int du \ \exp(-u^2 - a_n(xc^{-\frac{1}{2}} + m_n + u)^2 - a_n(xc^{-\frac{1}{2}} + m_n - u)^2)$$

$$\cdot (1 + \sigma_n(xc^{-\frac{1}{2}} + m_n + u)) \ (1 + \sigma_n(xc^{-\frac{1}{2}} + m_n - u))$$

$$= (1 + 2a_n)^{-\frac{1}{2}} \ \exp(-a_{n+1}(x + m_{n+1})^2) \tag{15.20}$$

$$\cdot \left(1 + 2\pi^{-\frac{1}{2}}(1+2a_n)^{\frac{1}{2}} \int du \ \exp(-(1+2a_n)u^2) \ \sigma_n(xc^{-\frac{1}{2}} + m_n + u)\right.$$

$$\left. + \pi^{-\frac{1}{2}}(1+2a_n)^{\frac{1}{2}} \int du \ \exp(-(1+2a_n)u^2) \ \sigma_n(xc^{-\frac{1}{2}} + m_n + u) \ \sigma_n(xc^{-\frac{1}{2}} + m_n - u)\right).$$

This is a sum of three terms. The first term is the new principal term. The second term, which is linear in $\sigma_{n,1}$ will be absorbed into $\sigma_{n+1,1}$, and the other term will make up part of $\sigma_{n+1,2}$. The bound (15.17) follows at once from hypercontractivity (Lemma 10.7). Upon substituting the bounds (15.18),(15.19) into (15.20), one gets the recursive relation for t_n. It remains to control the __remainder__ R_{n+1}. First of all, notice that

$$| (1 + \sigma_n (m_n + x)) \exp(-a_n (m_n + x)^2) |$$

$$\leq (1 + t_n + O(\| \sigma_{n,1} \|_{17,\gamma})) \exp(-b_n (m_n + x)^2) ,$$

and this will allow us to treat the crossed terms of the left hand and the right hand major contribution on the same footing as the terms of R_n. But the question of evaluating the action of $\mathcal{N}$ on the bound for R_n is purely a matter of Gaussian integrations. For $|k| \neq 2^j$, a suitable bound for the contribution of the term centered at $km_{n+1}2^{-j-1}$ to R_{n+1} is

$$s_n^2 \left((1+2a_n)/(1+2b_n) \right)^{\frac{1}{2}} (1 + t_{n+1} + O(\| \sigma_{n+1,1} \|_{17,\gamma}))$$

$$\cdot (1 + t_n + O(\| \sigma_{n,1} \|_{17,\gamma}))^{-2}$$

$$\cdot \sum_{\substack{p+q=k \\ |p|,|q| \leq 2^j}} \exp\left(L_n (p^2+q^2) - \frac{b_n}{2 + 4b_n} \frac{(p-q)^2}{2^{2j}} m_n^2 \right) .$$

The sum is then bounded as claimed in the statement of the lemma, since it is equal to

$$\sum_{r=|k|-2^{j+1}}^{2^{j+1}-|k|} \exp(L_n k^2/2) \exp\left((L_n/2 - b_n m_n^2/((2 + 4b_n) 2^{2j})) r^2 \right) .$$

This completes the proof of the lemma.

Next we "integrate" the recursion relations of Lemma 15.11. Let $n =$ $n_7 - n_4$. Then it is easy to see that for <u>Case 1</u>, i.e. for $L_{n+j} \leq$ $b_n m_n^2 / ((1 + 2b_n) 2^{2j})$, one has

$$L_{n+j} = (L_n + 4) 2^{-j} - 4 .$$

Next, still in Case 1, one finds

$$\log s_{n+j} = ((\log s_n) + O(1)) 2^j - O(1) + 4^{j+1} .$$

(Note that $s_n \sim \exp(-\varepsilon^{-1})$.) Now in <u>Case 2</u>, we have $L_{n+j} \geq$ $b_n m_n^2 / ((1 + 2b_n) 2^{2j})$, and this case takes place for $j \geq 1$, as one can check by substituting the initial values. Thus we find for $j > 1$,

$$L_{n+j} = b_{n+j} m_{n+j}^2 / ((1 + cb_{n+j}) 2^{2j}) - 2 ,$$

and this shows that the Case 2 persists for any bounded j provided ε is sufficiently small. We also find

$$\log s_{n+j} = -b_{n+j} m_{n+j}^2 / (1 + cb_{n+j}) + O(1) 2^j .$$

Now the stage is set for the <u>introduction of the measure</u>. We shall first rewrite the result of Lemma 15.11 for $j=27$ for the measure defined by $\exp(-x^2/2) \mathcal{N}^{n_7+27} (\phi_\varepsilon + f)(x)$. Let $n_8 = n_7 + 27$.

<u>THEOREM 15.12.</u> <u>For</u> n_8 <u>as defined above, the function</u> $\mu_{n_8}(x) =$ $\exp(m_{n_8 - n_4}^2 /2 - x^2/2) \mathcal{N}^{n_8} (\phi_\varepsilon + f)(x)$ <u>is proportional to a function</u> <u>of the form</u> (<u>with</u> $n = n_8 - n_4$)

$$\sum_\pm \exp(-A(m_n \pm x)^2 + m_n(m_n \pm x)) \cdot (1 + \sigma_n'((2A)^{\frac{1}{2}}(m_n \pm x))) + R_n'(x) ,$$

where $A = a_n + 1/2$ lies between 147 and 210 , and the bounds on σ'_n
and R'_n will be given in the proof.

Proof: Throughout, we omit the factor of proportionality in μ_{n_8} .
By Lemma 15.11, $\mu_{n_8} = \mathcal{N}^{n_8}(\phi_\varepsilon + f)$ is (for $n = n_8 - n_4$) of the form

$$\mu_{n_8}(x) = \sum_{\pm} \exp(-a_n(m_n \pm x)^2)(1 + \sigma_n(m_n \pm x)) + R_n(x).$$

A typical term of R_n is of the form

$$O(1) \exp\left\{(\delta^2 - 1) b_n m_n^2 / (1 + cb_n) - b_n(x + \delta m_n)^2\right\} , \qquad (15.21)$$

where $|\delta| \leq 1 - 2^{-27}$. If we multiply this by $\exp(m_n^2/2 - x^2/2)$, we
obtain

$$O(1) \exp\left(m_n^2\left\{(\delta^2 - 1) b_n/(1 + cb_n) - \delta^2 b_n/(1 + 2b_n) + 1/2\right\}\right)$$

$$\cdot \exp(-b'_n(x + \delta m'_n)^2) , \qquad (15.22)$$

where $b'_n = b_n + 1/2$, $m'_n = m_n b_n / b'_n$. Note that the coefficient of
the Gaussian is very small as a function of small ε provided

$$\delta^2 < 1 - (1 + cb_n)/(2(2-c)b_n^2) ,$$

which is satisfied for

$$|\delta| < 0,865 \qquad (15.23)$$

since $b_n > 5$. The terms for which (15.23) holds will be summed to define R'_n .

It thus remains to discuss the terms for which δ does not satisfy
(15.23) (but still $|\delta| < 1 - 2^{-27}$). These terms will be absorbed into
σ'_n (as well as σ_n) and the main point of having done the 27 steps was
to make sure that this operation gives only a small contribution.

We write $\sigma'_n = \sigma'_{n,1} + \sigma'_{n,2}$, where we shall absorb the remaining terms from R_n into $\sigma'_{n,2}$. Now it is easy to verify from

$$\sigma'_{n,1}((2A)^{\frac{1}{2}}(m_n \pm x)) \;=\; \sigma_{n,1}(m_n \pm x)$$

that we have a bound

$$\| \sigma'_{n,1} \|_{2,\frac{1}{4}} \;=\; O(1) \; \| \sigma_{n,1} \|_{17,\gamma} \quad .$$

Now $\sigma'_{n,2}((2A)^{\frac{1}{2}}(m_n + x)) = \sigma_{n,2}(m_n + x)$ + remaining terms on the left in R_n . We choose $b''_n = 5$ and we want to bound, in view of (15.19), the quantity

$$\sup_{1 \geq \delta \geq 0,865} \quad \sup_x \quad \exp\left((\delta^2 - 1) b_n m_n^2 / (1 + c b_n) - b_n (x + \delta m_n)^2 + b''_n (x + m_n)^2 \right) \quad ,$$

cf. (15.21). A straightforward calculation proves that $\sigma'_{n,2}$ is bounded by

$$|\sigma'_{n,2}((2A)^{\frac{1}{2}} z)| \quad \leq \quad O(\varepsilon^{1/8}) \; \exp((a_n - b''_n) z^2) \quad .$$

Remark: After the 27 steps, the number b_n satisfies $11 < b_n < 11,1$

The next proposition will allow us to do another k steps, where k is such that $(c/2)^k \simeq \varepsilon^{1/1000}$, while a variant of it will allow to make the (infinitely many) final steps. We shall prove recursively that the density

$$f_k(x) \;=\; \exp(-x^2/2) \; \mathcal{N}^{n_8 + k}(\phi_\varepsilon + f)(x)$$

has for $k \geq 0$, $(c/2)^k \geq \varepsilon^{1/22}$, the following properties.

P1) The density: It is of the form

$$f_k(x) = Q_k \left[\sum_{\pm} \exp(-A_k(M_k \pm x)^2/2) \exp(d_k(M_k \pm x)) \right.$$

$$\left. \cdot (1 + \sigma_k(A_k^{\frac{1}{2}}(M_k \pm x)))^{\cdot} + R_k'(x) \right] \quad ,$$

with $Q_k = O(1) \exp(Q' 2^k + Q'' c^k)$, and $Q' \neq 0$.

P2) The covariance: By Theorem 15.12 we have $A_o = 2A$. For $k > 0$, we have

$$A_k = A_o - 1 + (c/2)^k \quad ,$$

$$d_k = d_o (c/2^{\frac{1}{2}})^k \quad , \quad d_o = m_{n_8 - n_4} \quad .$$

Finally,

$$M_o = m_{n_8 - n_4} \quad ,$$

and for $k > 0$, one has

$$M_k = M_o 2^{k/2} \quad .$$

P3) The remainder σ_k: For $k = 0$, one has

$$\sigma_o = \sigma_o^{(0)} + \sigma_o^{(9)} \quad ,$$

where

$$\sigma_o^{(0)} = \sigma_{n,1}' \quad , \quad \sigma_o^{(9)} = \sigma_{n,2}' \quad , \quad n = n_8 - n_4 \ \text{(cf. Theorem 15.12)},$$

$$\| \sigma_o^{(0)} \|_{2,\frac{1}{4}} \leq O(\varepsilon^{1/8}) \quad , \quad \lambda_o^{(9)} = O(\varepsilon^{1/8}) \quad , \quad \lambda_o^{(j)} = 0, \ j=1,\ldots 8.$$

For $k > 0$, one has $\sigma_k = \sigma_k^{(0)} + \sigma_k^{(1)} + \ldots + \sigma_k^{(9)}$ with the following bounds. First for $\sigma_k^{(0)}$ one has

$$\| \sigma_k^{(0)} \|_{2,\rho_k} \leq 2(\rho_k A_k/4A_{k-1}\rho_{k-1}^3)^{\frac{1}{4}} \| \sigma_{k-1}^{(0)} \|_{2,\rho_{k-1}} + 100 \lambda_{k-1}^{(1)} \quad ,$$

where $\rho_k = 1/2 - (c/2)^k/4$, $\lambda_k^{(j)} \leq 100 \lambda_{k-1}^{(j+1)}$ for $0 < j < 9$.

Finally, we have

$$\lambda_k^{(9)} \leq 1000 \max \left((\lambda_{k-1}^{(j)})^2, \ \| \sigma_{k-1}^{(0)} \|_{2,\rho_{k-1}}^2 \ , \ \exp(-O(\epsilon^{-1}) 2^{0,01k}) \right).$$

For the other $\sigma_k^{(j)}$, with $j = 1,\ldots,9$ one has

$$| \sigma_k^{(j)}(z) | \leq \lambda_k^{(j)} \exp(b_k^{(j)} z^2/2) ,$$

$$b_k^{(1)} = 0,26 , \quad b_k^{(2)} = 0,41 \quad , \quad b_k^{(3)} = 0,58 \quad , \quad b_k^{(4)} = 0,73 ,$$
$$b_k^{(5)} = 0,84 , \quad b_k^{(6)} = 0,91 \quad , \quad b_k^{(7)} = 0,95 \quad , \quad b_k^{(8)} = 0,97 ,$$
$$b_k^{(9)} = 1 - B_k/A_k , \quad \text{see below for the definition of } B_k .$$

P4) The remainder R_k: Let $j = k + 27$. For $j \geq 27$, the function R_k is bounded as follows.

$$| R_k(x) | \leq \sum_{|p| \leq N_k} \exp\left(-\Lambda_o \ 2^{\alpha j} \ (1 - |p| 2^{-j})^{\alpha'} \right) \tag{15.24}$$

$$\cdot \exp\left(-B_k \ (x + pM_k' \ 2^{-j})^2 / 2\right) ,$$

where $\alpha = 1/2 - 2\epsilon$, $\alpha' = 1/2 - \epsilon$, $B_k = 10 + (c/2)^k$,

$$M_k' = 2^{k/2} M_o \ 10/B_k \ , \quad \Lambda_o = 0,05 M_o^2 \ 2^{-27\alpha} ,$$
$$N_k = 2^j - \text{integer part of} (\max\{2, 56 \ 10^{-3} \ 2^{27} \ 2^{(1-\beta)k} ; 0,135 \ 2^{27} \}),$$
$$\beta = (1-\alpha)/(2-\alpha') .$$

THEOREM 15.13. The rescaled density f_k defined by

$$f_k((2/c)^{k/2} x) = \exp(-x^2/2) \ \mathcal{N}^{n_8+k} (\phi_\epsilon + f)(x)$$

satisfies the relations P1 - P4 as long as $k \geq 0$ and $2^{-3k/2} < \epsilon^{1/8}$.

Proof: The case $k = 0$ follows by a straightforward but tedious inspection. The other cases ($k \rightarrow k+1$, $k \geq 0$) are handled by induction and we consider the same six terms as in Proposition 15.4.

Term R_1: From the definition of $\mathcal{N}$ it follows that the transformation of the density is

$$f \quad \to \quad \pi^{-\frac{1}{2}} \exp(\rho x^2 (c/2)^{k+1}) \int du\ f(2^{-\frac{1}{2}}x+u)\ f(2^{-\frac{1}{2}}x-u)\ , \quad (15.25)$$

where $\rho = (1/c - 1/2)$. Using this, we get at once that the term R_1 obtained from f_k is equal to

$$R_1(x) = A_k^{-\frac{1}{2}} \exp\left(\rho x^2(c/2)^{k+1} - A_k(x+2^{\frac{1}{2}}M_k)^2/2 + 2^{\frac{1}{2}} d_k (x+2^{\frac{1}{2}}M_k)\right)$$

$$\cdot \left(1 + \vartheta_{k+1}^{(1)}(A_{k+1}^{\frac{1}{2}} (x + 2^{\frac{1}{2}}M_k))\right) \Big\}\ ,$$

where

$$\vartheta_{k+1}^{(1)}\left((A_{k+1}/A_k)^{\frac{1}{2}} y\right) = \mathcal{A}_2(\sigma_k)(y) + \mathcal{N}_2(\sigma_k)(y)\ ,$$

and $\mathcal{A}_2$, $\mathcal{N}_2$ are our standard operators from Section 14. A straightforward Gaussian integration yields, using P3,

$$| \mathcal{N}_2(\sigma_k)(y)|$$

$$\leq \quad 10^3 \max \{ \|\sigma_k^{(0)}\|_{2,\rho_k}^2\ ,\ \lambda_k^{(i)^2}\ ,\ i=1,\ldots 8,\} \exp(b_k^{(9)} y^2/2)\ ,$$

and similarly

$$| \mathcal{A}_2(\sigma_k^{(i)})(y)| \quad \leq \quad 100 \exp(\delta_i' y^2/2) \lambda_k^{(i)}\ ,$$

where $\delta_i' = b_k^{(i)}/(2-b_k^{(i)})$, for $i = 1,\ldots,9$. We define now $\sigma_{k+1}^{(i)}$ for $i = 1,\ldots,8$ by

$$\sigma_{k+1}^{(i)}(y) = \mathcal{A}_2(\sigma_k^{(i+1)})\ ((A_k/A_{k+1})^{\frac{1}{2}} y)\ .$$

Then it follows that

$$|\sigma_{k+1}^{(i)}(y)| \leq \lambda_{k+1}^{(i)} \exp(b_{k+1}^{(i)} y^2/2) ,$$

for $i = 1,\ldots,8$ with $\lambda_{k+1}^{(i)} \leq 100 \lambda_k^{(i+1)}$, since $b_{k+1}^{(i)} > \delta_{i+1}' (A_k/A_{k-1})$, as one checks by a direct computation. For $i = 0$, we define

$$\sigma_{k+1}^{(0)}(y) = \mathcal{A}_2(\sigma_k^{(0)} + \sigma_k^{(1)})((A_k/A_{k+1})^{\frac{1}{2}} y) .$$

We have to bound $\| \sigma_{k+1}^{(0)} \|_{2,\rho_{k+1}}$. The term coming from $\sigma_k^{(1)}$ is bounded by

$$| \mathcal{A}_2(\sigma_k^{(1)})((A_k/A_{k+1})^{\frac{1}{2}} y)| \leq 100 \lambda_k^{(1)} \exp(0,15 \, y^2/2) .$$

The second term yields upon Gaussian integration, using $\rho_k A_k \leq \rho_{k+1} A_{k+1}$, the bound

$$\| \sigma_{k+1}^{(0)} \|_{2,\rho_{k+1}} \leq O(1) \lambda_k^{(1)} + (4\rho_{k+1}A_{k+1}/A_k\rho_k^3)^{\frac{1}{4}} \| \sigma_k^{(0)} \|_{2,\rho_k} ,$$

since $\| \mathcal{A}_2(f) \|_{2,\lambda}^2 \leq 2\lambda^{-1} \| f \|_{2,\lambda}^2$, if $\lambda < \frac{1}{2}$, as is verified from the definitions. Note that the products

$$2^{-k} \prod_{i=1}^{k} (4 \, \rho_{k+1} \, A_{k+1} / A_k \, \rho_k^3)^{\frac{1}{4}}$$

converge to a nonzero limit as $k \to \infty$. Inserting the various recursive definitions, we find that the term is

$$\exp(\rho (c/2)^{k+1} M_{k+1}^2) \cdot \sum_{\pm} \exp(-A_{k+1} (M_{k+1} \pm x)^2/2 + d_{k+1}(M_{k+1} \pm x)) ,$$

so that the factor Q_k in P1 satisfies the correct relations. In the further terms, we will have to split off the factor

$$\exp(\rho (c/2)^{k+1} M_{k+1}^2) .$$

Term R_2. This is the term in which $\mathcal{N}$, considered as a bilinear form acts on $\{R_k, R_k\}$, see also the beginning of the proof of Theorem 15.12. Let c_p be the factor $\exp(-\Lambda_o \, 2^{\alpha j} \, (1 - |p| 2^{-j})^{\alpha'})$ in the formula (15.24) for R_k. Then we have to bound

$$(R_k * R_k) (2^{-\frac{1}{2}} x) \cdot \exp(\rho \, (c/2)^{k+1} \, (x^2 - M_{k+1}^2) \,)$$

$$= \sum_{|p|,|q| \leq N_k} c_p \, c_q \, \exp(- B_k (x + (p+q) \, 2^{-j-1} \, 2^{\frac{1}{2}} \, M_k')^2/2)$$

$$\cdot \exp(\rho \, (c/2)^{k+1} \, (x^2 - M_{k+1}^2)) \, . \tag{15.26}$$

The contribution to the peak number s of R_{k+1} coming from (15.26) (let us call it c_s') is bounded by $c_s' \, \exp(- B_{k+1} \, (x+s2^{-j-1} M_{k+1}')^2/2)$ and

$$c_s' \leq \sum_{\substack{|p|,|q| \leq N_k \\ p+q = s}} c_p c_q \, \exp\big(2\rho (c/2)^{k+1} \, (s^2 2^{-2j-2} M_k'^2 \, B_k/B_{k+1} - M_k^2)\big) \, . \tag{15.27}$$

Substituting the recursive bounds on c_p, c_q, we find

$$c_s' \leq \sum_{\substack{|p|,|q| \leq N_k \\ p+q = s}} \exp\big(- \Lambda_o ((1 - |p| 2^{-j})^{\alpha'} + (1 - |q| 2^{-j})^{\alpha'})\big)$$

$$\cdot \exp\big(- \rho c^{k+1} \, M_o^2 \, (1 - 100(s2^{-j-1})^2 / B_k \, B_{k+1})\big) \, . \tag{15.28}$$

We distinguish two regions for s. In the first region, we have $N_{k+1} > |s| > 2^j$, and this implies that the p,q in (15.28) are of the same sign. Hence we find by the concavity of $(1-x)^{\alpha'}$ that

$$c_s' \leq (1/6) \, \exp\, (- \Lambda_o 2^{\alpha(j+1)} \, (1 - |s| 2^{-j-1})^{\alpha'}) \, , \tag{15.29}$$

provided ε is sufficiently small. In the second region, we have $0 \leq |s| \leq 2^j$, and we find from (15.28),

$$c_s' \leq 2^{2j+1} \, \exp(- \Lambda_o 2^{\alpha j}) \, \exp\big(- \rho \, c^{k+1} M_o^2 \, (1 - (s/2^{j+1})^2)\big). \tag{15.27'}$$

Since $t = |s/2^{2j+1}| \geq 1/2$, we find

$$(1 - t^2) \geq 2^{\alpha-1} (1-t)^{\alpha}, \tag{15,30}$$

and it is easy to check that for sufficiently small ε, (15.29) follows in the second case. Eq. (15.29) shows that the term R_2 can be absorbed into R_{k+1}.

Term R_6. This is the cross-term between the two main peaks. It only contributes to the term $s = 0$ of R_{k+1} and the coefficient it contributes is

$$2 \exp(d_k^2 / B_k) \ (1 + O(\varepsilon^{1/20})) \ \exp(- \rho(c/2)^{k+1} \ M_{k+1}^2)$$

$$= 2 \exp(d_k^2 / B_k) \ (1 + O(\varepsilon^{1/20})) \ \exp(- \rho \ c^{k+1} \ M_o^2) \ ,$$

taking into account the covariances of the terms $\sigma_k^{(j)}$. This is less than

$$(1/6) \ \exp(- \Lambda_o \ 2^{\alpha(j+1)}) \ ,$$

which is the necessary bound to absorb this term into R_{k+1} , provided

$$d_k^2 / B_k - \rho \ c^{k+1} \ M_o^2 \ < \ -\Lambda_o \ 2^{\alpha(j+1)} ,$$

which is easily seen to be satisfied. This shows that R_6 can be absorbed into R_{k+1}.

Term R_4. This is the cross-term between the main term and R_k. This term is the most difficult one in the present proof because it produces terms which cannot be absorbed into R_{k+1} alone but which must be absorbed into R_{k+1} and into the term with $\sigma_{k+1}^{(9)}$. This mechanism would fail in the case without interaction and in that case these

seemingly "uninteresting and negligeable" error terms would lead to a
final convergence to one Gaussian, since we know that there is no
phase transition in the free case. The difficulty with this term is
therefore not just a technical point but reflects in what a subtle way
the interaction is involved in the formation of the phases.

First of all, from the recursive bounds we get

$$|R_4(x)| \leq \sum_{|p| \leq N_k} \exp\left(\rho \, (c/2)^{k+1} \, (x^2 - M_{k+1}^2)\right)$$

$$\cdot c_p \exp(d_k^2 / 2B_k) \, (1 + O(\varepsilon^{1/20}))$$

$$\cdot \exp\left(- B_k \, (x + (2^j + p) 2^{-j-1} \, 2^{\frac{1}{2}} \, M_k')^2 / 2\right) . \tag{15.31}$$

The term number p gives thus a contribution to a Gaussian centered at
$(2^j + p) \, 2^{-j-1} \, M_{k+1}'$, and as p varies over the set $|p| \leq N_k$, this set
will exceed N_{k+1} . In the cases of interest, one has $\mu 2^{\beta - 1} <$
$1 - (2^j + p)/2^{j+1} < \mu$, where $\mu = 2,56 \cdot 10^{-3}$. For those p for which
$|2^j + p| \leq N_{k+1}$, we may absorb the contribution into R_{k+1} . This is
done as follows. Let $s = 2^j + p$. We have to show that the contribution
from (15.31) for this value of p is (say) 1/6 of the bound given for
the term number s in R_{k+1} . We do only the argument for $s \geq 2^j$, the
other case being very similar to the treatment of R_3. Omitting the factor $1 + O(\varepsilon^{1/20})$ for convenience, we have to show as in the transition from (15.27) to (15.28) ,

$$3 \, c_{s-2^j} \, \exp\left(d_k^2/2B_k - \rho c^{k+1} \, M_0^2 \, (1 - 100 \, (s2^{-j-1})^2/B_k B_{k+1})\right)$$

$$\leq \frac{1}{6} \exp\left(-\Lambda_0 \, 2^{\alpha(j+1)} \, (1 - |s| 2^{-j-1})^{\alpha'}\right) . \tag{15.32}$$

It follows from the definitions that the first exponential in (15.32) equals

$$\exp(d^2_{k+1}/2B_{k+1} - d^2_k/2B_k) \exp(-\rho\, c^{k+1}\, M^2_o\, (2s'-s'^2)) \,,$$

where $s' = 1 - |s|/2^{j+1}$. Since we have assumed $s \geq 2^j$, we find that

$c_{s-2^j} \leq \exp(-\Lambda_o\, 2^{\alpha(j+1)}\, s'^{\alpha'})$, as before, cf. (15.30). Therefore it suffices to show

$$d^2_{k+1}/2B_{k+1} - d^2_k/2B_k - \rho\, c^{k+1}\, M^2_o\, (2s' - s'^2) < u < 0 \,. \qquad (15.33)$$

This is easily seen to be satisfied with $u = M^2_o\, (c/2)^k\, (0,005-0,27\rho c)$,

since $s' \gtrsim 0,135\ 2^{-k}$, as can be seen from the definition of N_k in property P4. We now come to the case of those p in (15.32) for which

$|2^j + p| > N_{k+1}$. From (15.32) we find that a term with a given $s = 2^j + p > N_{k+1}$ is bounded by

$$\Gamma F(x)$$

$$= \Gamma\, \exp(-\Lambda_o\, 2^{\alpha(j+1)}\, (1-\delta)^{\alpha'})\ \exp(-B_{k+1}\, (x + \delta M'_{k+1})^2 / 2) \,, \qquad (15.34)$$

where $\delta = s/2^{j+1}$ and $|\Gamma| \leq \exp\left(M^2_o\, (c/2)^k\, (0,005-0,27\rho c)\right)$ and also

$|\Gamma| \leq \exp(2^{k/6}\, M^2_o\, m)$ for $k > 200$ and with $m < 0$. These bounds are derived using the definition of N_k and Eq.(15.33) . We shall bound the function $F(x)$ by

$$G(x)\ +\ H(x)$$

$$= \exp(d^2_{k+1} / 2B_{k+1} - B_{k+1}\, (x + M'_{k+1})^2 / 2)$$

$$+ \exp(-\Lambda_o\, 2^{\alpha(j+1)}\, (1-\delta')^{\alpha'})\ \exp(-B_{k+1}\, (x + M'_{k+1}\, \delta')^2 / 2) \,,$$

with $\delta' < \delta$. First we determine δ' by intersecting the graphs of F and G. It is easy to check that

i) $\delta' < \delta$, $\delta'\, 2^{j+1} \leq N_{k+1}$,

ii) $F(-\delta' M'_{k+1}) = G(-\delta' M'_{k+1}) < H(-\delta' M'_{k+1})$,

so that the situation is as depicted in Fig. 10.

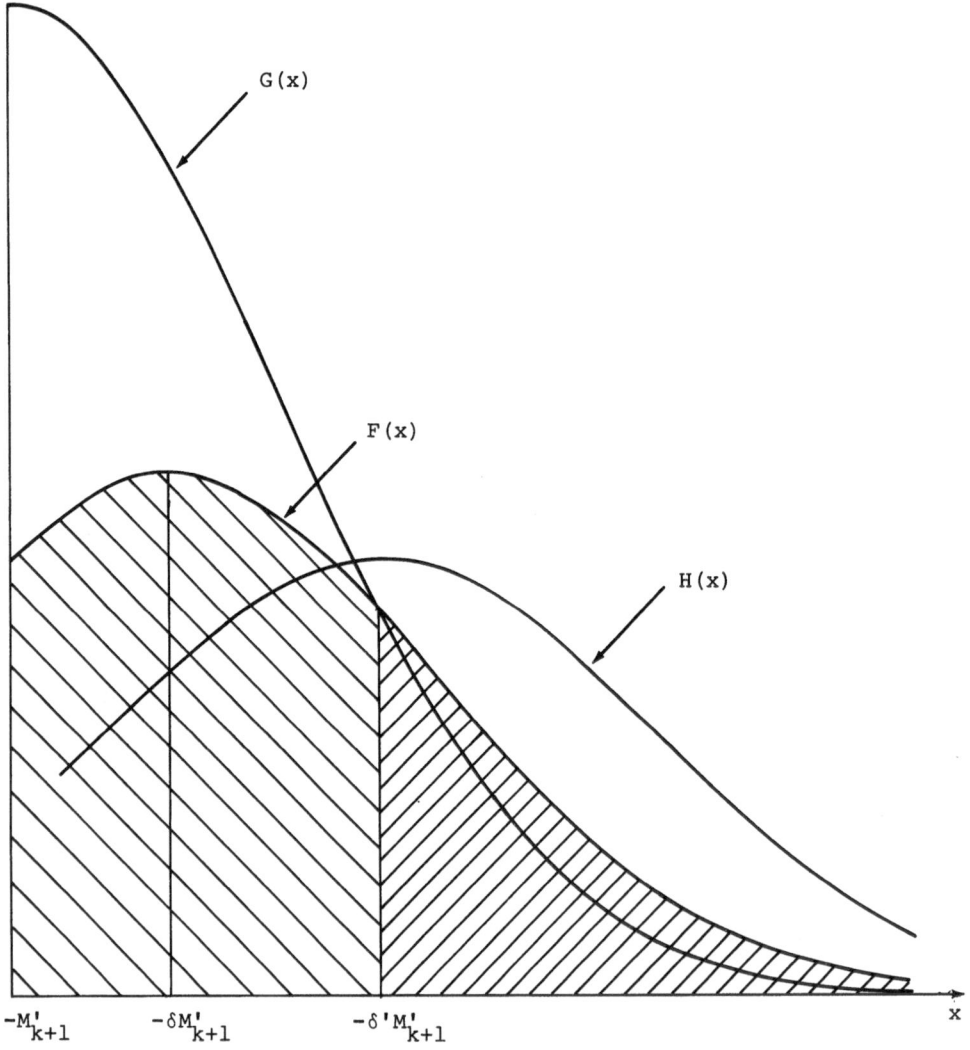

$-M'_{k+1}$ $-\delta M'_{k+1}$ $-\delta' M'_{k+1}$

Figure 10. Absorption of the difficult term.

Since Γ is very small, we see that the terms can be absorbed as fol-

lows: We shall absorb the right hand part of the contribution into

R_{k+1}, more precisely into the peak centered at (about) $-\delta' M'_{k+1}$, while the left hand part will be absorbed into

$$\exp(\rho\,(c/2)^{k+1}\,M^2_{k+1})\,\exp\!\left(-A_{k+1}\,(M_{k+1}+x)^2/2 + d_{k+1}\,(M_{k+1}+x)\right)$$

$$\cdot\sigma^{(9)}_{k+1}(\,A^{\frac12}_{k+1}\,(M_{k+1}+x))\quad,$$

or into its bound

$$\exp(d^2_{k+1}/\,2B_{k+1} - B_{k+1}(M'_{k+1}+x)^2/\,2)\cdot\lambda^{(9)}_{k+1}\quad.$$

The induction step and hence the proof of the theorem is complete.

We next state a lemma which is a variant of Theorem 15.13 and which suffices to prove the final convergence. We use the preceding theorem as long as $k < k_1$, where k_1 is defined by $2^{k_1} < \varepsilon^{-1/11}$.

LEMMA 15.14. The rescaled density $f_k((2/c)^{k/2}x) = \exp(-x^2/2) \cdot \mathcal{N}^{n_8 +k}(\,\phi_\varepsilon + f)(x)$ satisfies for $k \geq k_1$ the relation

$$f_k(x) = Q_k\left[\;\sum_\pm \exp\!\left(-A_k(M_k\pm x)^2/2 + d_k(M_k\pm x)\right)\right.$$

$$\left.\cdot\left(1 + \sigma_k(A^{\frac12}_k\,(M_k\pm x))\right) + R_k(x)\;\right]\quad,$$

with $d_k = (c/2)^k M_k$. The remainder R_k satisfies P4, and

$$\|\,\sigma^{(0)}_k\,\|_{2,\rho_k}\;\leq\;D_k\;=\;(1,01\cdot 2^{-\frac12})^{k-k_1}\,O(\varepsilon^{3/88})\quad,$$

where $\rho_k = \frac12 - \frac14(c/2)^k$. As a new feature, $\sigma^{(0)}_k((2\rho_k)^{-\frac12}x)$ is orthogonal to 1, x, x^2 in $L_{2,\frac12}$, and this implies slightly different recursion relations for A_k and M_k , namely

$$A_{k+1} = A_k - 2\,(c/2)^{k+1} + O((c/2)^k)\;,\quad M_{k+1} = 2^{\frac12}M_k + O(D_k\,(c/2)^k).$$

The other relations are as in Theorem 15.13.

Proof: The initial case $k = k_1$ is easily checked from Theorem 15.13. The iteration steps are the same in all cases except for the term R_1. The treatment of the term R_1 is similar in spirit to the case of Theorem 14.6, except that now we have also a linear term (magnetization). We fix $k \geq k_1$ and we consider the term R_1' which we define as R_1 without the term which is quadratic in σ,

$$R_1'(x) = \exp\left(\rho(c/2) \ x^2 - A_k(x + 2^{\frac{1}{2}}M_k)^2 + d_k 2^{\frac{1}{2}}(x + 2^{\frac{1}{2}}M_k)\right)$$

$$\cdot \pi^{-\frac{1}{2}} \int du \ \exp(-A_k u^2) \ \left(1 + 2\sigma_k (A_k^{\frac{1}{2}}(2^{-\frac{1}{2}}x + M_k + u))\right).$$

Note that

$$R_1'(x) = \exp\left(-A_k' \ (x + 2^{\frac{1}{2}}M_k)^2 + 2\rho(c/2)^{k+1}M_k^2 + d_k'(x + 2^{\frac{1}{2}}M_k)\right)$$

$$\cdot A_k^{-\frac{1}{2}} \ L_k(A_k^{\frac{1}{2}}(x + 2^{\frac{1}{2}}M_k)) \quad ,$$

where $A_k' = A_k - 2\rho(c/2)^{k+1}$, $d_k' = 2^{\frac{1}{2}}d_k - 2^{3/2}\rho(c/2)^{k+1} M_k$,

$$L_k(x) \quad = \quad 1 + \mathcal{A}_2(\sigma_k)(x) \ .$$

The norm of $\mathcal{A}_2$ on $L_{2,\frac{1}{2}}$ is 2 which is not good for our purpose and this is the reason for projecting $\sigma_{k+1}^{(0)}$ onto the complement of 1, x, x^2 in $L_{2,\frac{1}{2}}$ and absorbing the rest into the constants A_{k+1} and M_{k+1} . We write

$$L_k \quad = \quad 1 + \sum_{i=0}^{9} \mathcal{A}_2(\sigma_k^{(i)})$$

$$= \quad 1 + \sigma_{k+1}^{(0)}{}' + \left(\mathcal{A}_2(\sigma_k^{(0)} + \sigma_k^{(1)}) - \sigma_{k+1}^{(0)}{}'\right) + \sum_{i=1}^{8} \sigma_{k+1}^{(i)} \quad ,$$

where

$$\sigma_{k+1}^{(0)}{}'(x) \quad = \quad \mathcal{A}_2((\sigma_k^{(0)} + \sigma_k^{(1)}) \circ (2\rho_k)^{-\frac{1}{2}}) ((2\rho_k)^{\frac{1}{2}}x) \quad ,$$

and

$$\sigma_{k+1}^{(i)}(x) = \mathcal{A}_2(\sigma_k^{(i+1)}) \quad , \text{ for } i = 1,\ldots,8 \ .$$

The difference

$$\delta\sigma_k = \mathcal{A}_2(\sigma_k^{(0)} + \sigma_k^{(1)}) - \sigma_{k+1}^{(0)}{}'$$

will be absorbed into $\sigma_{k+1}^{(9)}$. Finally, we define $\sigma_{k+1}^{(0)}$ through the condition: $\sigma_{k+1}^{(0)}((2\rho_k)^{-\frac{1}{2}} x)$ is orthogonal to 1, x, x^2 in $L_{2,\frac{1}{2}}$ and satisfies the equation

$$K \exp\left(- A_{k+1} (x + M_{k+1})^2/2 + d_{k+1}(x + M_{k+1})\right)\left(1 + \sigma_{k+1}^{(0)}(A_{k+1}^{\frac{1}{2}}(x+M_{k+1}))\right)$$

$$= \exp\left(- A_k'(x + M_{k+1})^2/2 + d_k'(x + 2^{\frac{1}{2}}M_k)\right)\left(1 + \sigma_{k+1}^{(0)}{}'(A_k^{\frac{1}{2}}(x+2^{\frac{1}{2}}M_k))\right) \ .$$

$$(15.35)$$

The proof of the lemma is complete if we check three things:

i) $\quad |\delta\sigma_k|(x) \ \leq\ D_k\ (c/2)^k \exp(b_{k+1}^{(9)}\ x^2/2) \quad ,$

ii) $\quad \|\sigma_{k+1}^{(0)}\|_{2,\rho_{k+1}} \ \leq\ D_{k+1} \quad ,$

iii) $\quad A_k' - A_{k+1} = O((c/2)^k) \quad , \quad 2^{\frac{1}{2}}M_k - M_{k+1} = O(D_k^2) \ .$

Indeed, the condition i) above implies that

$$\lambda_{k+1}^{(9)} \ \leq\ 2000 \cdot \max(\ D_k^2\ ,\ \lambda_k^{(i)2}\ ) \quad ,$$

which is slightly different from the bound of Theorem 15.13, but still amply sufficient for the recursion relations. The proof of i) follows from the inequalities

$$\int du \ \exp(-u^2) \ |f(\alpha^{-1}(2^{-\frac{1}{2}}\alpha z + u)) - f(2^{-\frac{1}{2}}z + u)|$$

$$= \int du \ \exp(-u^2) \ |f(2^{-\frac{1}{2}}z + u)| \cdot |\alpha \exp(-(\alpha-1)u^2) - 1| \ \leq$$

$$\leq\ O(\alpha-1)\ \|\ f\ \|_{2,\frac{1}{2}}\ \exp(x^2/3)\ .$$

To prove ii) and iii) , one writes the definition of $\sigma_{k+1}^{(0)}$ using (15.35) , with the unknowns A_{k+1} , M_{k+1} , K, and then one writes the orthogonality relations for $\sigma_{k+1}^{(0)}$. Note that $d_{k+1} - d_k = (c/2)^{k+1}(M_{k+1} - 2^{\frac{1}{2}}M_k)$. The equations to be solved in the unknowns $\beta = M_{k+1} - 2^{\frac{1}{2}}M_k$, $\alpha = A_{k+1} - A_k'$ and K are

$$\int dy\ y^p\ \exp(-\rho_{k+1}y^2)$$

$$\cdot\Big(1\ -\ K\ \exp\big(\alpha y^2/2A_{k+1} + y(\beta(c/2)^{k+1} + \beta A_k' - \alpha\beta)A_{k+1}^{-\frac{1}{2}}\big)$$

$$\cdot\ \exp\big(-\alpha\beta^2/2 - \beta^2(c/2)^{k+1} - \beta^2 A_k' - \alpha\beta^2\big) \tag{15.36}$$

$$\cdot\big(1 + \sigma_{k+1}^{(0)'}(A_k^{\frac{1}{2}}(A_{k+1}^{-\frac{1}{2}}y - \beta))\big)\Big)\ =\ 0\ ,\ \text{for}\ p = 0,1,2.$$

It is now a straighforward but tedious matter to check that the equations (15.36) have a unique solution near $\alpha = 0$, $\beta = 0$, $K = 1$, and that the bounds ii) and iii) hold, cf. also the proof of Theorem 14.6. This concludes the proof of Lemma 15.14.

As a consequence of Lemma 15.14, we have now the final result.

THEOREM 15.15. Let $\phi_\varepsilon + f \notin \mathcal{W}_s$, $\|\ f\ \|_\infty \leq \varepsilon^{330}$, $\varepsilon > 0$ sufficiently small, $\phi_\varepsilon + f > 0$ and even. Suppose further that f is on the side of positive coefficients for $e_2 \sim +(2x^2 - 1)\ \exp(-\varepsilon\theta x^4)$. Then $\mathcal{N}^n(\phi_\varepsilon + f)$ converges to two δ-functions "like a Gaussian" in the following sense: There is a finite, non-zero constant μ such that

i) The limit
$$\lim_{n\to\infty}\ \frac{1}{2^n}\ \log\ \int dx\ \exp(-x^2/2)\ \mathcal{N}^n(\phi_\varepsilon + f)(x)$$
exists.

ii) <u>One can decompose</u> $\mathcal{N}^n(\phi_\varepsilon + f)(x) = g_n(\mu c^{n/2} + x) + g_n(\mu c^{n/2} - x)$

<u>in such a way that the following limits exist and are different</u>

<u>from zero.</u>

$$\lim_{n \to \infty} \frac{(2/c)^{np} \int ds \exp(-s^2/2 - \mu c^{n/2} s) g_n(s) s^{2p}}{\int ds \exp(-s^2/2 - \mu c^{n/2} s) g_n(s)} .$$

<u>They are the moments of a Gaussian measure.</u>

<u>Proof</u>: The statement i) follows trivially from Lemma 15.14 and the definition of Q_k. To prove ii), a possible definition of g_n is as follows. Recall that

$$f_k((2/c)^{k/2} x) = \exp(-x^2/2) \mathcal{N}^{n_8+k}(\phi_\varepsilon + f)(x).$$

By Lemma 15.14, we have therefore

$$\mathcal{N}^{n_8+k}(\phi_\varepsilon + f)(x)$$

$$= \text{const.} \left[\sum_\pm \exp\left(-(A_k - (c/2)^k)(M_k \pm (2/c)^{k/2}x)^2/2\right) \right.$$

$$\left. \cdot \left(1 + \sigma_k(A_k^{\frac{1}{2}}(M_k \pm (2/c)^{k/2}x))\right) + \frac{1}{2} R_k(\pm x) \right]$$

$$= \left[\mathcal{N}^{n_8+k}(\phi_\varepsilon + f)\right]_+ (+x) + \left[\mathcal{N}^{n_8+k}(\phi_\varepsilon + f)\right]_+ (-x) ,$$

and we define therefore

$$\mu = c^{-n_8/2} \lim_{k \to \infty} M_k/2^{k/2} , \quad \alpha = \lim_{k \to \infty} A_k ,$$

and

$$g_n(\mu c^{n/2} + x) = \left[\mathcal{N}^n(\phi_\varepsilon + f)\right]_+ (x) .$$

We find thus with $k = n - n_8$,

$$g_n(s)$$

$$= \text{const.} \left[\exp\left(-(2/c)^k (A_k - (c/2)^k) (s + c^{k/2} (M_k 2^{-k/2} - c^{n_8/2} \mu))^2 / 2\right) \right.$$

$$\cdot (1 + \sigma_k (A_k^{\frac{1}{2}} (s + c^{k/2} (M_k 2^{-k/2} - c^{n_8/2} \mu))))$$

$$\left. + \frac{1}{2} R_k (x - c^{n/2} \mu) \right] .$$

To prove the convergence in the sense of ii) we consider

$$\frac{\displaystyle\int_{-\infty}^{\infty} ds \; g_n((c/2)^{-k/2} s) \; \exp(-(c^2/2)^k \mu \, c^{n_8/2} s) \; F(s)}{\displaystyle\int_{-\infty}^{\infty} ds \; g_n((c/2)^{-k/2} s) \; \exp(-(c^2/2)^k \mu \, c^{n_8/2} s)} \; .$$

Now the error terms containing σ_k and R_k tend to zero, provided $F(s)$ is polynomially bounded, as is easily seen from the bounds on $\|\sigma_k\|$ in Lemma 15.14 and on R_k in Theorem 15.13, Property P4. This completes the proof of Theorem 15.15.

16. Miscellaneous Short Proofs

Proof of Lemma 5.1 :

1) We show there are positive functions on $\mathcal{V}_s$. The point of the proof is that the codimension of the stable manifold is 2, while we construct a 3-parameter family of positive functions in a neighborhood of ϕ_ε .

Proof : Let C be the cube in $\mathbb{R}^3$ defined by $C = \{(\alpha,\beta,\gamma) \mid |\alpha|,|\beta|,|\gamma| < \varepsilon^{200}\}$. Then $(\alpha,\beta,\gamma) \rightarrow T(\alpha,\beta,\gamma)(x) = \phi_\varepsilon(x) \exp\left(\alpha + \beta\, e_2(x) + \beta\, e_4(x)\right)$ is a C^4 diffeomorphism of C onto a subset of L_∞ , and this subset is contained in the domain of the nonlinear "diagonalizing" operator S . Therefore $S \circ (T - \phi_\varepsilon)$ is a C^4 diffeomorphism of C onto a subset of L_∞ , and $S \circ (T-\phi_\varepsilon)(0,0,0) = 0$. In the decomposition $L_\infty = E_u \oplus E_s$ we have

$$S \circ (T - \phi_\varepsilon)(\alpha,\beta,\gamma) = \left(x(\alpha,\beta,\gamma), y(\alpha,\beta,\gamma)\right)$$

where $x(\cdot)$ and $y(\cdot)$ are C^4 functions from C to some subsets of E_u and E_s respectively, and $x(0,0,0) = 0$, $y(0,0,0) = 0$. $x(\cdot)$ may now be considered as a C^4 function from a neighborhood of zero in $\mathbb{R}^3$ into a neighborhood of zero in $E_u = \mathbb{R}^2$, therefore there exists a number $a < \varepsilon^{200}/3$ such that in a ball of radius a the rank of $x(\cdot)$ is not less than the rank of $x(\cdot)$ in $(0,0,0)$ [40]. Now we have $D\left(S \circ (T - \phi_\varepsilon)\right)(0,0,0) = DS_{(0)} \circ D(T - \phi_\varepsilon)(0,0,0)$. But $DS_{(0)} = I$ and $D(T - \phi_\varepsilon)(0,0,0)(\delta\alpha,\delta\beta,\delta\gamma) = \delta\alpha\phi_\varepsilon + \delta\beta\phi_\varepsilon e_2 + \delta\gamma\phi_\varepsilon e_4$. From $\phi_\varepsilon e_2 = e_2 + O(\varepsilon)e_4 + O(\varepsilon^{\frac{1}{2}})$ and $\phi_\varepsilon e_4 = e_4 + O(\varepsilon)e_6 + O(\varepsilon^2)e_8 + O(\varepsilon^{\frac{1}{2}})$ we have rank $x(\cdot)_{(0,0,0)} = 2$ since $Dx(\delta\alpha,\delta\beta,\delta\gamma) = (\delta\alpha,\delta\beta) + O(\varepsilon^{\frac{1}{2}})(|\delta\alpha|+|\delta\beta|+ |\delta\gamma|)$. Now we can apply the implicit function theorem [29] to $x(\cdot)$ in a ball of radius $a' < a$ sufficiently small since $\dfrac{D\left(S \circ (T - \phi_\varepsilon)\right)}{D(\alpha,\beta)}(0,0,0)$ is a diffeomorphism from $B_{\mathbb{R}^3}(0,a')$ to a neighborhood of zero in $\mathbb{R}^2$.

Therefore we have for a strictly positive number $a'' < a'$ the existence of a curve $\gamma \to (\alpha(\gamma), \beta(\gamma))$ defined if $|\gamma| < a''$ and such that $x(\alpha(\gamma), \beta(\gamma), \gamma) = 0$. Then $S \circ (T - \phi_\varepsilon)(\alpha(\gamma), \beta(\gamma), \gamma) = (0, y(\alpha(\gamma), \beta(\gamma), \gamma))$, and since S^{-1} is a diffeomorphism, $\phi_\varepsilon + (T - \phi_\varepsilon)(\alpha(\gamma), \beta(\gamma), \gamma)$ is on the stable manifold and by construction a positive function, different from ϕ_ε for $\gamma \neq 0$.

2) The other assertions of Lemma 5.2 follow at once from the results on ϕ_ε, Sections 9 - 11.

THEOREM 16.1.

For fixed $\varepsilon \geq 0$ there are constants K_n such that

$$| \partial_x \phi_\varepsilon(x) | \leq K_n \phi_\varepsilon^{\frac{1}{2}}(x)$$

for all $x \in \mathbb{R}$.

Proof : Let $\mathfrak{X}_\alpha = \{ f \in L_\infty | \exists K \in \mathbb{R}^+ , |f(x)| \leq K \phi_\varepsilon^\alpha(x) \ \forall x \in \mathbb{R} \}, \alpha > 0$. With the norm $\|f\|_\alpha = \sup_{x \in \mathbb{R}} |f(x)| / \phi_\varepsilon^\alpha(x)$, $\mathfrak{X}_\alpha$ is a Banach space.

LEMMA 16.2. $\mathcal{A}_{\phi_\varepsilon}$ is a bounded and compact operator in $\mathfrak{X}_{\frac{1}{2}}$.

Proof : If $\|f\|_{\frac{1}{2}} \leq 1$ then

$$| \mathcal{A}_{\phi_\varepsilon}(f)(x) | \leq 2\pi^{-\frac{1}{2}} \|f\|_{\frac{1}{2}} \int \exp(-u^2) \phi_\varepsilon(xc^{-\frac{1}{2}} - u) \phi_\varepsilon^{\frac{1}{2}}(xc^{-\frac{1}{2}} + u) \, du$$

$$\leq 2 \int \exp(-u^2/2) \pi^{-\frac{1}{4}} \phi_\varepsilon^{\frac{1}{2}}(xc^{-\frac{1}{2}} - u) \left[\pi^{-\frac{1}{4}} \exp(-u^2/2) \phi_\varepsilon^{\frac{1}{2}}(xc^{-\frac{1}{2}} - u) \right.$$
$$\left. \cdot \ \phi_\varepsilon^{\frac{1}{2}}(xc^{-\frac{1}{2}} + u) \right] du$$

$$\leq 2[1 + 0(\varepsilon)] \left[\pi^{-\frac{1}{2}} \int \exp(-u^2) \phi_\varepsilon(xc^{-\frac{1}{2}} - u) \phi_\varepsilon(xc^{-\frac{1}{2}} + u) \, du \right]^{\frac{1}{2}}$$
$$\cdot \left[\pi^{-\frac{1}{2}} \int \exp(-u^2) \phi_\varepsilon(xc^{-\frac{1}{2}} - u) \, du \right]^{\frac{1}{2}}$$

$$\leq \ell(x) \, \phi_\varepsilon^{\frac{1}{2}}(x)$$

where $\ell(x) = 2[1 + O(\varepsilon)] \, \pi^{-\frac{1}{2}} \int \exp(-u^2) \phi_\varepsilon(xc^{-\frac{1}{2}} - u) \, du$. We see that if $x \to \pm \infty$ then $\ell(x) \to 0$, so that $\phi_\varepsilon^{\frac{1}{2}}(x) \, \mathcal{A}_{\phi_\varepsilon}(f)(x)$ is an equicontinuous family of functions which goes uniformly to zero if $x \to \pm \infty$ when f varies in the unit ball of $\mathfrak{X}_{\frac{1}{2}}$. By the theorem of Ascoli-Arzela, $\mathcal{A}_{\phi_\varepsilon}(\cdot)$ is a compact operator of $\mathfrak{X}_{\frac{1}{2}}$.

COROLLARY 16.3.

The spectra of $\mathcal{A}_{\phi_\varepsilon}\big|_{L_\infty}$ and $\mathcal{A}_{\phi_\varepsilon}\big|_{\mathfrak{X}_{\frac{1}{2}}}$ coincide except possibly for $\{0\}$.

Proof : By the Riesz-Schauder theory and the topological inclusion $\mathfrak{X}_{\frac{1}{2}} \subset L_\infty$ and $\mathfrak{X}_{\frac{1}{2}} \subset L_{2,\gamma}$ we have $\mathrm{Sp}_{\mathfrak{X}_{\frac{1}{2}}} \mathcal{A}_{\phi_\varepsilon} \subset \mathrm{Sp}_{L_{2,\gamma}} \mathcal{A}_{\phi_\varepsilon} = \mathrm{Sp}_{L'_{2,\gamma}} \mathcal{A}^*_{\phi_\varepsilon} \cup \{0\}$

$$\subset \mathrm{Sp}_{\mathfrak{X}'_{\frac{1}{2}}} \mathcal{A}^*_{\phi_\varepsilon} \cup \{0\} = \mathrm{Sp}_{\mathfrak{X}_{\frac{1}{2}}} \mathcal{A}_{\phi_\varepsilon} \cup \{0\} .$$

We return to the proof of Theorem 16.1. From Lemma 16.2 and Corollary 16.3 it follows that since the first few eigenvalues of $\mathcal{A}_{\phi_\varepsilon}$ are simple, the corresponding eigenfunctions are in $\mathfrak{X}_{\frac{1}{2}}$. But the eigenfunction e_3 equals ϕ_ε' (by Proposition 10.2), so that $\phi_\varepsilon' \in \mathfrak{X}_{\frac{1}{2}}$. Now we proceed recursively as in Section 10, using the inequality that if $f, g \in \mathfrak{X}_{\frac{1}{2}}$ then

$$x \to \int \exp(-u^2)(1 + |u|) \, f\,(x/\sqrt{c} + u) \, g\,(x/\sqrt{c} - u) \, du$$

is also in $\mathfrak{X}_{\frac{1}{2}}$. This proves Theorem 16.1.

COROLLARY 16.4.

The assertions of Theorem 16.1 hold for functions of the form $\phi(x) = \phi_\varepsilon(x) \exp(-a - b\,e_2 - c\,e_4)$ with $|a| + |b| + |c| < \varepsilon^{100}$.

The results of page 195 and the Corollary 16.4 prove the existence of functions satisfying $(c_1) - (c_5)$ of Section 5 and hence the Lemma 5.1 is proved.

Proof of Lemma 5.2 : The claim is: $\beta \to \phi(\beta, \cdot)$ is C^2 in L_∞ and its derivative $\partial_\beta \phi(\beta, \cdot)\big|_{\beta=\alpha} = \Phi_2$ satisfies $P_{e_0} \Phi_2 \neq 0$ where P_{e_0} is the spectral projection associated with λ_0, provided that ϕ satisfies the hypothesis of Lemma 5.1, and $\alpha \neq 2\pi e^{-1}(2 - \sqrt{2})$.

Proof : We have

$$\phi(\beta,z) \;=\; \exp\!\big((\beta/\alpha)\log \phi \,(z(\alpha/\beta)^{\frac{1}{2}})\big)\left[\frac{4\pi}{c}\left(\frac{2-c}{\alpha c}\right)^{\frac{1}{2}}\right]^{-(\beta/2\alpha - 1/2)}$$

and it is sufficient to prove that $t \to f(t) = \exp\!\big(t \log \phi \,(z\, t^{-\frac{1}{2}})\big)$ is C^2 in L_∞ near $t = 1$. Now from Lemma 5.1

$$t \to g_1(t) \;=\; \exp\!\big(t \log \phi(z\, t^{-\frac{1}{2}})\big)\left[\log \phi(z\, t^{-\frac{1}{2}}) - \tfrac{1}{2}tz\,\frac{\phi'(zt^{-\frac{1}{2}})}{\phi(zt^{-\frac{1}{2}})}\right]$$

is continuous in L_∞ if $|t-1| < \tfrac{1}{4}$ and therefore $\displaystyle\int_1^t g_1(\tau)\,d\tau + \phi(z)$ which is pointwise equal to f is in L_∞ .

In the same way

$$t \to g_2(t) \;=\; \exp\!\big(t \log \phi(zt^{-\frac{1}{2}})\big)\left[\log \phi(z\, t^{-\frac{1}{2}}) - \frac{tz}{2}\left[\frac{\phi'(zt^{-\frac{1}{2}})}{\phi(zt^{-\frac{1}{2}})}\right]\right]^2$$

$$+ \exp\!\big(t \log \phi(zt^{-\frac{1}{2}})\big)\left[-\frac{tz}{2}\left(\frac{\phi'(zt^{-\frac{1}{2}})}{\phi(zt^{-\frac{1}{2}})}\right) - \frac{1}{2}\frac{z\phi'(zt^{-\frac{1}{2}})}{\phi(zt^{-\frac{1}{2}})}\right.$$

$$\left. + \frac{1}{4}t^2 z^2\,\frac{\phi''(zt^{-\frac{1}{2}})}{\phi(zt^{-\frac{1}{2}})} - \frac{1}{4}t^2 z^2\,\frac{(\phi'(zt^{-\frac{1}{2}}))^2}{\phi^2(zt^{-\frac{1}{2}})}\right]$$

is continuous in L_∞ if $|t-1| < \tfrac{1}{4}$, and therefore $\displaystyle\int_1^t g_2(\tau)\,d\tau + g_1(1)$

which is pointwise equal to $g_1(t)$ is in L_∞ since $g_1(1)$ is.

A direct computation now gives

$$\partial_\beta \phi(\alpha,z) = -\frac{1}{2\alpha}\left\{1 + \log\left|\frac{4\pi}{c}\left(\frac{2-c}{\alpha c}\right)\right|\right\}\phi(z) + \frac{1}{\alpha}\left\{\phi(z)\log\phi(z) - \frac{z}{2}\partial_z\phi(z)\right\}.$$

By perturbation theory in L_∞, we have from Lemma 5.1

$$\partial_{\dot\beta}\phi(\alpha,z) = -\frac{1}{2\alpha}\left\{1 + \log\left|\frac{4\pi}{c}\left(\frac{2-c}{\alpha c}\right)\right|\right\} + O(\varepsilon)$$

which is different from zero if $\alpha \neq 2\pi\, e^{-1}\,(2-\sqrt{2})$ and ε is sufficiently small.

Reference:

[40]   J. DIEUDONNÉ : Foundations of Modern Analysis, New York, 1968,

Academic Press. (§10.3)

Topics in Applied Physics

Founded by **Helmut K. V. Lotsch**

This book series is devoted to research achievements of current interest. Each volume deals with a different topic under the editorship of a recognized authority in the field. It covers application-oriented aspects of the topic under consideration, the basic physical principles being summarized in a comprehensive introduction.
The contributors to each volume are internationally known experts. The publication periods are comparable with those of scientific journals to keep pace with the rapidly accumulating results.

Springer-Verlag
Berlin
Heidelberg
New York

Selected Issues from
Lecture Notes in Mathematics

Lecture Notes in Physics